ENERGY RESEARCH AT THE CUTTING EDGE

ENERGY RESEARCH AT THE CUTTING EDGE

JOHN R. BLACKWOOD (EDITOR)

Nova Science Publishers, Inc.
New York

Senior Editors: Susan Boriotti and Donna Dennis
Coordinating Editor: Tatiana Shohov
Office Manager: Annette Hellinger
Graphics: Wanda Serrano
Editorial Production: Vladimir Klestov, Matthew Kozlowski and Maya Columbus
Circulation: Ave Maria Gonzalez, Vera Popovich, Luis Aviles, Raymond Davis,
Melissa Diaz and Jeannie Pappas
Communications and Acquisitions: Serge P. Shohov
Marketing: Cathy DeGregory

Library of Congress Cataloging-in-Publication Data
Available Upon Request

ISBN: 1-59033-480-9

400 Oser Ave, Suite 1600
Hauppauge, New York 11788-3619
Tele. 631-231-7269 Fax 631-231-8175
e-mail: Novascience@earthlink.net
Web Site: http://www.novapublishers.com

Printed in the United States of America

CONTENTS

PREFACE

Energy fuels modern society, though it is often taken for granted. The issues surrounding energy supply of paramount importance to governments the world over. While calls for "clean" energy technologies are rampant and loud, conventional sources like coal and oil remain the most feasible. However, research continues to plow ahead to find better and more efficient ways to keep the demand for energy met. Nuclear power is a perennially controversial but enticing possibility, although many nations seek to phase-out nuclear power, citing safety and environmental concerns. The energy industry and production is a major economic component in any country, so all energy and electricity issues hold sway over us all.

This book brings together an important selection of studies focusing on the energy industry, its current state and future. Included are looks at nuclear power and economic policies to influence the electric industry. This current research combines for a needed resource in the study of energy.

Chapter 1

SUPPLY FUNCTION EQUILIBRIA UNDER ALTERNATIVE CONDITIONS WITH DATA FROM THE POLISH ELECTRICITY MARKET

Elzbieta Kuba[1] and Friedel Bolle[2]

ABSTRACT

The paper has two simple messages. (1) In particular because of the low elasticity of demand for electricity, the industry can be far from complete competition even with large numbers of suppliers. Thus, industry policy should carefully examine mergers and acquisitions by competitors. (2) Because of the low slope of demand curves as well as the low slope of equilibrium supply curves, stochastic demand reductions lead to large price reductions. Requiring the suppliers to submit bids for a long time (a day, a week), while demand bids are short term (an hour or half an hour), will considerably reduce prices (for a given industry structure).

I. INTRODUCTION

In July, 2000, the Polish spot market for electricity was launched. This was a milestone in the liberalisation and restructuring of the Polish electricity industry but not its completion. At the moment and in the years coming, the privatisation of producers, of local distribution companies, and of the grid company PSE is the most important issue. As in England, Scandinavia, and in Germany, in this process of privatisation and liberalisation the industry structure will be changed. At the moment, there are 27 national producers of electricity, but in the long run much less are expected to exist.

While, in the beginning, the British electricity system was the model structure for the Polish reforms, later on the Scandinavian market inherited this function. The reform process

[1] RWE Rheinbraun AG, Köln

[2] Europa-Universität Viadrina, Frankfurt (Oder)

was admittedly slow and cumbersome, but one should keep in mind that the transformation of the electricity industry did not take place in a liberal market economy as in Western Europe but in a country where the whole economic system was under reconstruction.

Spot market prices are an indicator of the present value of electricity. They serve to directly evaluate futures contracts (which are intrinsically contracts for differences), and they also evaluate the long-term contracts which currently dominate the relations between the producers and retailers/industrial consumers.

In the following analysis, we make some simplifying assumptions:

(i) The complete supply and (wholesale) demand of electricity is traded on the spot market. Long term contracts and futures are ignored.

This is far from true, at least at the moment. In the long run, however, 20 – 40 % of the production might be traded on the spot market. We do not think that the adjustment of our model to such a restricted share will influence the results essentially. A futures market (and also long-term contracts) are expected to have a dampening influence on spot market prices (see Allaz, 1992; Powell, 1993; Bolle, 1993); but again we do not think that the qualitative results of our model are severely effected[3].

(ii) There will be a relatively small number of producers (10 or 20 in our model) who can influence the spot price and, therefore, act strategically. There will be a relatively large number of consumers who do not believe that they can influence the price and therefore can be modelled by a systematically and stochastically varying demand function.

Ten strategically acting producers are still a lot and in many industries this number would be assumed to be close to complete competition. It would be assumed that there is no big difference between 10 and 20 – but this is not true for the market we are going to model and under our assumption of equilibrium selection. Thus we did not investigate even smaller numbers of "big players", say 3 to 6[4]; this assumption of 10 to 20 producers provides us with the additional message that under certain circumstances oligopolistic tendencies are still to be found for large numbers of players.

(iii) While the cost function of the producers is estimated (with a bit of guess-work) from Polish data, the linear demand function is adjusted to the Polish market only with respect to its absolute level. The slope is assumed to take the lowest value of Green and Newbery's (1992) interval[5] of possible slopes of the British demand.

It is assumed that short-term demand in Poland will still be less responsive than that in Britain 10 years ago. Demand reduction because of high spot prices is due to demand reductions of industrial users of electricity and retailers, mostly because they have own

[3] As the price in oligoploy models is severely effected by the level of demand we will discuss the problem in more depth in the appendix.

[4] Dowzycki (2002) expects 2 or 3 international and 5 or 6 polish electricity companies to emerge from the privitization process.

[5] Which was also relatively arbitrarily chosen.

production facilities; in the case of industrial users also availability of other energies or buffer stocks for products might cause demand reduction. All this, however, is expected to be relevant in Poland to a rather restricted extent which makes the demand function rather flat and more dependant on random (or systematic daily or weekly) influences than on price changes.

(iv) Market equilibrium (with identical producers) can be described by a supply function equilibrium. The equilibrium selection device (for the continuum of equilibria) is *symmetry* and *Pareto optimality* from the viewpoint of the producers (who are the only strategically acting players in this game).

While, from a theoretical point of view, these criteria of equilibrium selection are satisfactory, in real markets decisions might be difficult with so many equilibria. But as the same players participate again and again in the same game they might at least coordinate on that equilibrium which is the most profitable one for all of them – if they do not go even beyond this point and collude with non-equilibrium strategies.

The main results of this investigation are

- The number of producers has considerable influence on the market price, even for numbers between 10 and 20.
- The time interval, for which price offers are submitted, matters a lot with respect to prices under low demand conditions; the larger this time interval, the lower the prices.

The results also shed some light on the British spot market. The early market conditions with only two “relevant” producers, offered far-reaching cooperation opportunities. Recently, the old market structure has been substituted by a Scandinavian-type market, i.e. there is more emphasis on bilaterally negotiated long-term contracts, and there is a spot market with hourly instead of daily offers by the producers. As already said above, more long-term (as well as) futures contracts have a dampening effect on spot prices; the transition from daily to hourly offers by the producers has the opposite effect.

In the next section, the spot market for electricity in Poland is described. In Section III, the theory of supply function equilibria is briefly reconsidered. In Section IV, we show how cost and demand functions for Poland are estimated and, in Section V, the results of III and IV are put together for a simulation of the future Polish spot market. Section VI contains conclusions.

II. The Polish Electricity Market

The Polish electrictiy industry as well as the whole Polish economy is in a process of restructuring. The present state is governed partly by heritages from communist days while on the other hand important elements of a market system are already present – but even so still not working satisfactory.

Polish electricity prices of households and industry are rather low (see Table 1), the household prices have less than half the level of German prices and they are lower than in all other EC-countries – but this is no indication of low costs and competitiveness in the production of electricity. One indication of this is that Poland has a trade deficit in electricity with Germany; another indication are spot prices in Poland and Germany: In 2000, the average spot market price was about 100 PLN/MWh in Warsaw which was not much different from wholesale prices in Germany.

Table 1: Electricity prices in Poland and some Western European Countries (Jan. 1, 2002)

	1	2	3	4	5	6
Poland	10,93	8,94	6,84	6,19	5,05	4,50
Germany	29,23	17,29	9,03	8,11	7,67	5,98
England & Wales	19,11	9,80	6,80	6,01	6,28	5,66
Norway	41,79	41,79	5,11	5,11	5,10	5,10

Source: Power in Europe 369, 11 February 2002.
Prices in Eurocent/kWh
1 = Domestic Standard tariff 500 kWh annual consumption
2 = Domestic Standard tarif 7,500 kWh annual consumption
3 = Industrial tariff 500 kW annual maximum demand/1,752 MWh annual consumption
4 = Industrial tariff 500 kW annual maximum demand/3,504 MWh annual consumption
5 = Industrial tariff 2,500 kW annual maximum demand/8,760 MWh annual consumption
6 = Industrial tariff 2,500 kW annual maximum demand/17,520 MWh annual consumption

Though the price of electricity increased considerably (even with respect to average income it is now more than twice as high as in 1990), it is still “too low”, a burden from communist days when electricity was purposefully subsidised. The normalization process is still under way, and it is necessary because only prices which reflect (private and social) costs can be the basis of a rational industrial as well as environmental policy[6].

So consumer prices have to rise in order to provide the industry with the means to improve the efficiency of Polish power plants as well as their environmental performance. Some progress has been made in the last decade, but there is still a lot to do if Poland wants to meet the environmental standards of Western Europe.

On the other hand prices should not be so high that producers earn large oligopoly profits. In England and Wales, the liberalized market induced large improvements in efficiency and large cost reductions. For a long time, however, it seemed that the electricity industry profited much more than electricity consumers. While prices slightly increased (sligthly decreased in real terms) from 1990 till 1995, the profits of the Regional Electricity Companies and the profits of National Power and PowerGen doubled. Nuclear Electric earned 1994/95 a profit of one billion £ instead of a loss of nearly the same amount in 1989/90. So, the aim is to create a market with enough incentives for improving efficiency but without opportunities to earn undue profits. In one word: the market should be competitive.

Restructuring plans for the Polish electricity industry reach back as far as to 1990. The process was admittedly slow and difficult[7], but at last several important steps on the way to a competitive market system have been made.

[6] This is not only important for Poland but also for other countries as, for example, Germany where energy taxes and subsidies often contradict environmental goals.

[7] See Kuba, 2001

The first big step was the vertical disintegration and restructuring of the industry. There are now 27 producers of electricity (plus some industrial production), 33 local distribution companies[8], and an independent grid company, PSE, which manages the high voltage grid and coordinates the physical delivery of electricity. PSE is party in the old long-term contracts of the producers and distribution companies, and it has its own trading affiliate which admistrates these contracts. Since June 30, 2000, there is an electricity exchange in Warsaw, where hourly day-ahead contracts are traded. In September 2001 a balance market (also called regulatory market) for short-time adjustments of supply and demand in the whole Polish electricity system was launched.

The next big step is the privatisation of producers and distribution companies, and perhaps the restructuring of PSE who should concentrate on its transmission and coordination business. The distribution business of the local distribution companies is planned to be separated from trading and to be comprehended in 5 or 6 pure transportation companies. This should facilitate the opening of the market for smaller customers.

Presently, the distribution companies get about 60 % of their demand on the basis of old[9] long-term contracts, the majority of their rest demand is covered by new short-term (one to two years) bilateral contracts and only very small quantities are bought on the spot market. During the first half year since its launching about 300 MW per hour, i. e. 1 – 2 % of the wholesale demand, were traded on the spot market. Meanwhile the spot market has even less weight. Instead of this, the balancing market quickly gained importance and now, in some periods, amounts to 10 % of the whole market trade. However, conditions in Poland are still not "normal". After the privatisation (and that means to a great deal: internationalisation) of producers, local distribution companies, and also some big industrial consumers, the climate for the spot-market may become better and trade may reach volumes as in Norway[10] (20 – 40 %) or in England and Wales (10 % - 40 %) under the old conditions of mandatory participation in the spot market[11]. The German power exchanges LPX and EEX cover 6 – 7 % of the market volume but their long-term goal (also after merger) is to reach 20 %.

III. Supply Function Equilibria

Supply function equilibria were first investigated by Klemperer and Meyer (1989) and Bolle (1989, 1991); Green and Newbery (1992) simulated the British spot market on this basis, Bolle (2001) investigated equilibria with supply and demand functions.

The alternative theoretical approach to such models is a multiple object auction (Engelbrecht-Wiggans, 1988; Bolle, 1997; Ausubel and Crampton, 1998). A hybrid supply function/auction model has been proposed by von der Fehr and Harbord (1998).

---

[8] Some of which closely cooperate, for example the group G8, and may merge even before privatization.

[9] Concluded in 1994 till 1998 with time horizons up to 25 years. There are plans to transform (and renegotiate) these high price contracts into contracts for differences. This would require, however, a liquid spot market.

[10] The number depends a bit on how you count. Statnett, the Norwegian transmission company, estimates that some 40 % of consumption is covered by long-term contracts while the rest is bought on the spot market. Source: http://www.eurelectric.org/docs/2001-030-0351-12.pdf

[11] 10 % to 40 % = (100% mandatory spot market) minus 60 % to 90 % long term contracts). Source: Competition Commission (2000). Under the new market regime the voluntary spot market covers 2 – 3 % total market volume.

In principle, both model types can be applied in every case. However, practically all applications of supply function equilibria are to be found in the area of spot markets for electricity, while multiple bid auctions are applied to such fields as central bank auctions (Nautz, 1998) or spectrum auctions (Ausubel and Crampton, 1998; Bolle and Breitmoser, 2002). An exception is the empirical investigation of Wolfram (1998) who investigated the British spot market on the basis of an auction model. The central difference between these approaches are not the continuous (functions) vs. discrete (bids) instruments, but the typical accompanying assumptions in these models. One difference is that supply function equilibria are investigated, up to now, only under a "single price" rule, while auctions are also investigated under "you pay what you bid" or the Vickrey rule. Another difference is: Auctions are traditionally investigated under the assumption of complete information or of private or common values, typically the competitors' costs or values are private information. Supply function equilibria under complete information do not provide us with meaningful results, all existing models assume uncertainty with respect to demand; the competitors' costs are assumed to be common knowledge. Under not too complicate cost and demand functions, equilibria can be explicitly discussed. So, when deciding which model type to use in an oligopoly situation, one should ask[12] what pricing rule is applied and where the crucial uncertainty is to be found: is it market demand or is it competitors' costs?

The computation of equilibrium strategies requires, in both model types, the solving differential equations. Thus we get one or more free parameters or, in other words, we get a continuum of equilibria. In such a situation, equilibrium selection is crucial. We come back to this point below.

In electricity markets, competitors are rather well informed about costs. On the other hand, demand can differ a lot, randomly (weather, TV program, etc) as well as systematically (time of day, day of week)[13]. So, we decided to establish a supply function model with the following assumptions:

- There are many consumers of electricity whose demand is represented by a stochastic demand function $D(p,n)$ where p represents the price and n a random variable. So we assume that the participants in the spot market who represent demand do not think that they have market power.
- Bids can be approximated by a twice differentiable function: Producer i supplies $q_i = S_i(p)$, $i = 1,\ldots,k$;. The aggregate production equals demand:

$$\sum_{i=1}^{k} q_i = D(p,n),$$

- Production costs are represented by a linear or quadratic function $C_i(q_i)$. The linear term in this function is assumed to be 0, i.e. we regard prices p net of this linear term.

[12] Provided you would like to use existing theoretical investigations.

[13] The source of this variation is not important for the derivation of equilibria. The longer the time period for which bids (supply functions) must be submitted, the more important are systematic variances.

The supply functions are the strategic instruments of the producers who want to maximise their profits

$$(1)\quad \Pi_i = \int_{\underline{n}}^{\overline{n}} \left(S_i(p)p - C_i(S_i(p))\right) f(n)dn, \text{ for all } i,$$

where $f(n)$ is the density function of the demand parameter $n \in (\underline{n}, \overline{n})$.

Assuming a linear demand function

$$(2)\quad D(p,n) = n - mp, \text{ with } m = const. > 0,$$

a linear cost function (zero marginal costs when prices are net of these costs), and taking into account

$$(3)\quad S_i(p) = D(p,n) - \sum_{i \neq j} S_j(p),$$

the profit of producer i is

$$(4)\quad \Pi_i = \int_{\underline{n}}^{\overline{n}} \left(n - mp - \sum_{j \neq i} S_j(p) \right) pf(n)dn \text{ - fix costs.}$$

Now imagine that producer i regards his optimisation problem as the determination of the optimal price (for every n). This requires that the first order condition

$$(5)\quad \left[\left(-m - \sum_{j \neq i} S_j'(p) \right) p + \left(n - mp - \sum_{j \neq i} S_j(p) \right) \right] f(n) = 0$$

as well as the second order condition (for $f(n) > 0$)

$$(6)\quad -2m - 2\sum_{j \neq i} S_j'(p) - \sum_{j \neq i} S_j''(p) < 0$$

be fulfilled. In the symmetrical case, i.e. for $S_j = S$ for all j (and for $f(n) \neq 0$), (5) implies:

$$(7)\quad S(p) = (m + (k-1)S'(p))p \quad \text{or}$$

$$(8)\quad S'(p) = \frac{S(p)}{p(k-1)} - \frac{m}{(k-1)}.$$

The linear differential equation (8) is solved by:

$$(9)\quad S(p,a) = \begin{cases} ap - mp \cdot \ln(p) & for\ k = 2 \\ ap^{\frac{1}{k-1}} - \dfrac{mp}{k-2} & for\ k > 2, \end{cases}$$

a is an arbitrary real number from the interval $[a_{\min}, \infty]$. All these functions are concave, they increase for small p and they decrease for large p. The maximum value of S is to be found at

$$(10)\quad p = \begin{cases} e^{\frac{a}{m}-1} & for\ k = 2 \\ \left(\dfrac{m}{a} \cdot \dfrac{k-1}{k-2}\right)^{(k-1)/(2-k)} & for\ k > 2, \end{cases}$$

For prices smaller than this p, the second order condition is fulfilled. It is fulfilled even for a certain range of values beyond this p (see Bolle, 1991, and Figure 1), but it is not possible to use decreasing supply functions in an auction market where the auctioneer orders the bids and thus produces only non-decreasing supply functions. With increasing a (increasing supply) and given demand the spot price decreases. So, as long as the price is below the monopoly price, all producers are interested in small a. This provides us with a natural.

Equilibrium selection criterium: The symmetric equilibrium with the smallest possible $a = a_{min}$ is chosen.

a_{min} is determined by the requirement that, in all market situations, the market price p is connected with non-decreasing supply functions of the producers. So, the demand function with the maximal parameter $n = \overline{n}$ has to cut the aggregate supply function in its maximum. (10) implies

$$(11)\quad a_{mim} = m(1 + \ln p_{\max}) \text{ for k = 2}$$

$$(12)\quad a_{mim} = m\frac{k-1}{k-2} p_{\max}^{\frac{k-2}{k-1}} \text{ for k > 2}$$

where the market price in the case of maximal demand is

(13) $p_{\max} = \frac{\bar{n}}{m} \cdot \frac{1}{k+1}$.

The latter equation can be derived by setting $S'(p_{\max}) = 0$ in (7) which implies

(14) $S(p_{\max}) = mp_{\max}$,

and by taking into account that kS has to cut the (maximal) demand curve.

Figure 1 shows us the possible supply function equilibria and the related market prices - also for cases with more general cost functions. Aggregate supply can be as high as indicated by the marginal cost supply (complete competition), which is connected with $a=\infty$; the lowest possible supply is connected with a_{min}, i.e. the aggregate supply curve OA.

In the case of a quadratic cost function

(15) $C(S) = const + \frac{1}{2}cS^2$

the differential equation for the symmetric equilibrium supply functions is no longer linear.

(16) $(k-1)S'(p) = \frac{S(p)}{p - cS(p)} - m$

can be solved by parameterisation (see Kuba, 2001). We get

(17) $p_{\max} = \bar{n} \cdot \frac{mc+1}{km + m(mc+1)}$

(18) $S(p_{\max}) = \frac{m}{mc+1} \cdot p_{\max}$

For the computation of the selected S(p) we rely on numerical solutions of (16) which pass the point (p_{max}, $S(p_{max})$).

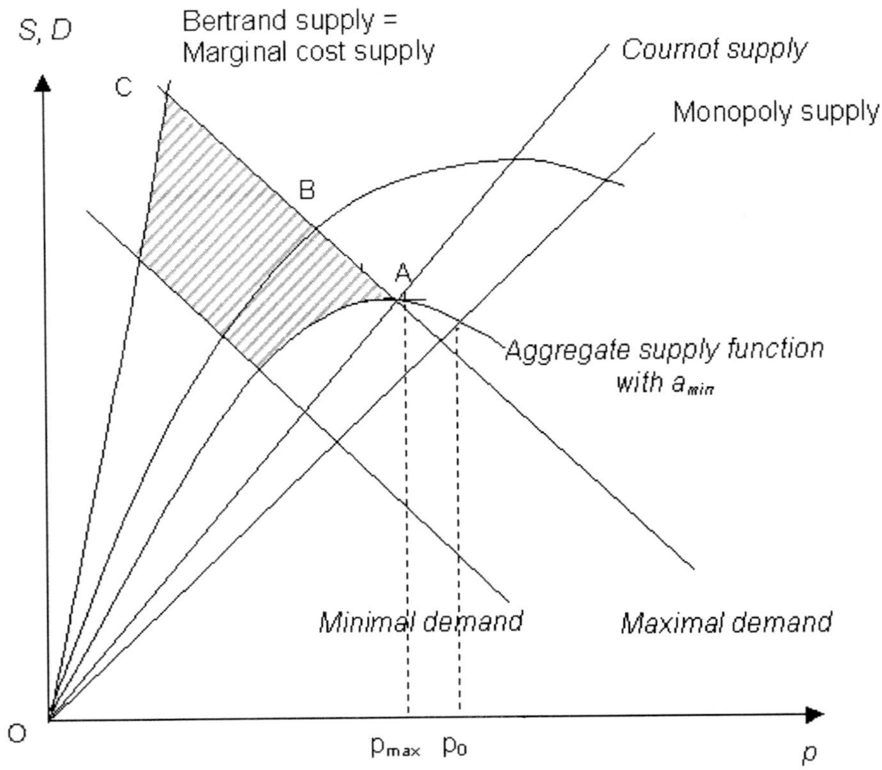

Figure 1: Equilibrium supply functions and the area of possible market prices (shaded) when an auctioneer orders the bids. The second order conditions are fulfilled for prices below p_0.

IV. Guessing and Estimating Costs and Demand of Polish Electricity

Currently, there are 27 producers in Poland. During the privatisation process, however, this number will certainly shrink. We want to assume only a mild concentration process because oligopolistic tendencies are nonetheless considerable. Our alternative assumptions are k= 10 and k = 20.

IV.1 Costs

The average marginal costs of electricity production in Poland were 48.14 PLN/MWh in 1998[14]. This value has to be added to prices determined by the simple model "without costs". It must also be considered when determining the demand functions. As, however, for the determination of a quadratic cost function we will use only the lignite and hard coal power

[14] Average marginal costs per MWh were computed from ARE (1999 a), Tabela 8, p. 10 and Tabela 1, p. 6.

plants of the 27 producers, and as we want to compare the respective results, we will carry out our simulation with the average marginal cost of these plants which are 58.26 PLN/MWh (see below).

We now want to estimate a quadratic cost function which takes into account that there are different marginal costs due to different fuels and different types of power plants. Because of missing data we take into account only plants which burn lignite or hard coal (which produce 90% of the gross production of electricity)

Power plants burning lignite – 8360 MW of installed capacity - had average marginal costs of 47.88 PLN/MWh[15] in 1998. With respect to the availability of this capacity, two classes of plants could be identified[16]. We assume this difference as accompanied by a difference in marginal costs. We assume that the more efficient plants – 2600 MW – have marginal costs of 46.5 PLN/MWh and the less efficient – 5760 MW – 48.5 PLN/MWh[17]. For hard coal plants - 13480 MW – three groups of plants were identified[18]. The total avarage costs of 64.71 PLN/MWh[19] were differentiated, considering a most efficient group (6360 MW) with average costs of 64 PLN/MWh, a medium group (5680 MW) with 65 PLN/MWh and a least efficient group (1440 MW) with 66.71 PLN/MWh.

In the case of a simulation with k identical producers we provide each of them with 1/k of the capacity of the classes identified, i.e. for k=10 (see Figure 2):

- Lignite power plants (836 MW) with MC=47.88 PLN/MWh:
  - 1st class (260 MW) with MC=46.5 PLN/MWh
  - 2nd class (576 MW) with MC=48.5 PLN/MWh
- Hard coal power plants (1348 MW) with MC=64.71 PLN/MWh
  - 1st class (636 MW) with MC=64 PLN/MWh
  - 2nd class (568 MW) with MC=65 PLN/MWh
  - 3rd class (144 MW) with MC=66.71 PLN/MWh.

For 10 producers, the least square estimation of a linear marginal cost function results in

$$(19)\ MC = 0{,}011\,\mathrm{S(p)} + 45{,}41[PLN/MWh].$$

For 20 producers, we get individual marginal cost functions

$$(20)\ MC = 0{,}022 \cdot S(p) + 45{,}41\ \text{[PLN/MWh]}.$$

[15] ARE (1999 a), Tabela 8.1, p. 10.

[16] In the first class of lignite PP up to 88 % of installed capacity could be used for power generation, in the second – less then 88 % were available, see ARE (1999 b).

[17] This is not more than a guess, taking into consideration also the opinion of Polish experts. The same applies for the marginal costs of the three categories of hard coal plants. The times of usage of power plants (which served to form the classes) were taken from ARE (1999b).

[18] The three classes of hard coal PP are as follows: I. Class: availability of more than 90 %, II. Class: between 90 % and 84 %; III. Class: 83 % and less, see ARE (1999 b).

[19] ARE (1999 a), Tabela 8.2, p. 11.

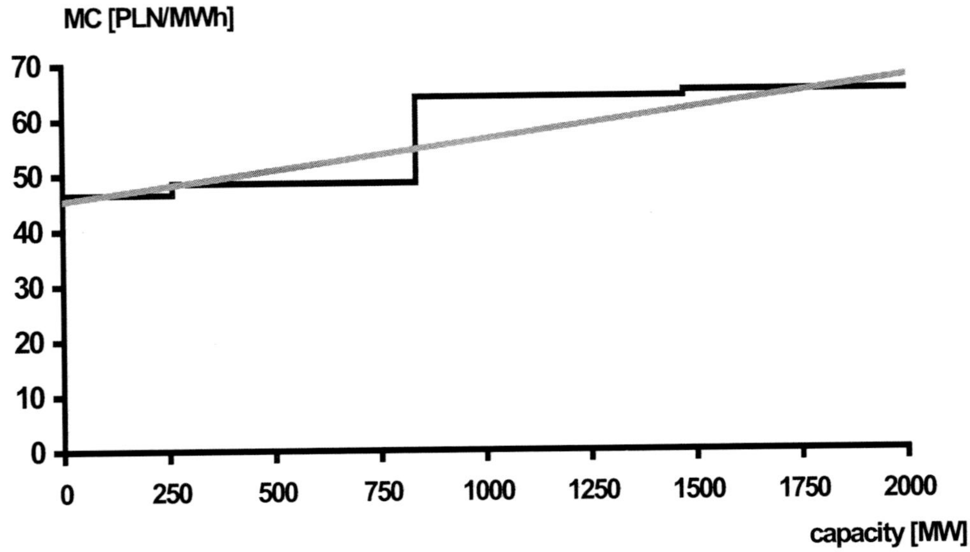

Figure 2: Marginal costs (MC) in the case of 10 producers.

IV.2 Demand

As far as we know, there are no empirical investigations concerning the wholesale demand of electricity in Poland. This is also a principal problem because electricity prices under the old regime practically did not vary. So, how should we get the values m and $\bar{n}$ of the (maximal) demand function? Green and Newbery (1992) chose (in a similar situation for England und Wales) relatively arbitrarily three possible values of m, namely $m = 100$ (MW)^{2}h/£, $m = 250$ (MW)^{2}h/£ and $m = 500$ (MW)^{2}h/£, with the medium value as the most plausible one[20]. We will use the smallest of the three values because we think that Polish retailers and industry will be even less flexible[21] than their British counterparts. As, however, electricity costs play a major role in some industries[22] we can assume that there is some flexibility and that there will be some investment in improved flexibility. So $m = 100$ (MWh)^2/£ (or $m = 15{,}72$ (MWh)^2/PLN[23]) seems to be an appropriate value for Poland (where S measures demand for one hour). The influence of m on maximal prices is to be seen in (13) and (17). For small c = dMC/dS, doubling m is accompanied by half as large p_{max}. $\bar{n}$, on the other hand, is easier to estimate.

Figure 3 shows two daily load curves from 1998, one shows the largest (Dec. 3, 1998) and the other shows the smallest demand (June 19, 1998).

---

[20] Green and Newbery (1992), p. 944, Green (1996), p. 210.

[21] Flexibility mainly means that demand can be postponed or supply can be substituted by own production.

[22] An extreme example is the Californian energy crises where aluminium producers shut down their facilities because it was more profitable to resell the electricity they had ordered under long term contracts.

[23] The rate of exchange on Dec. 31, 2000: 1 PLN= 0,1572 GBP.

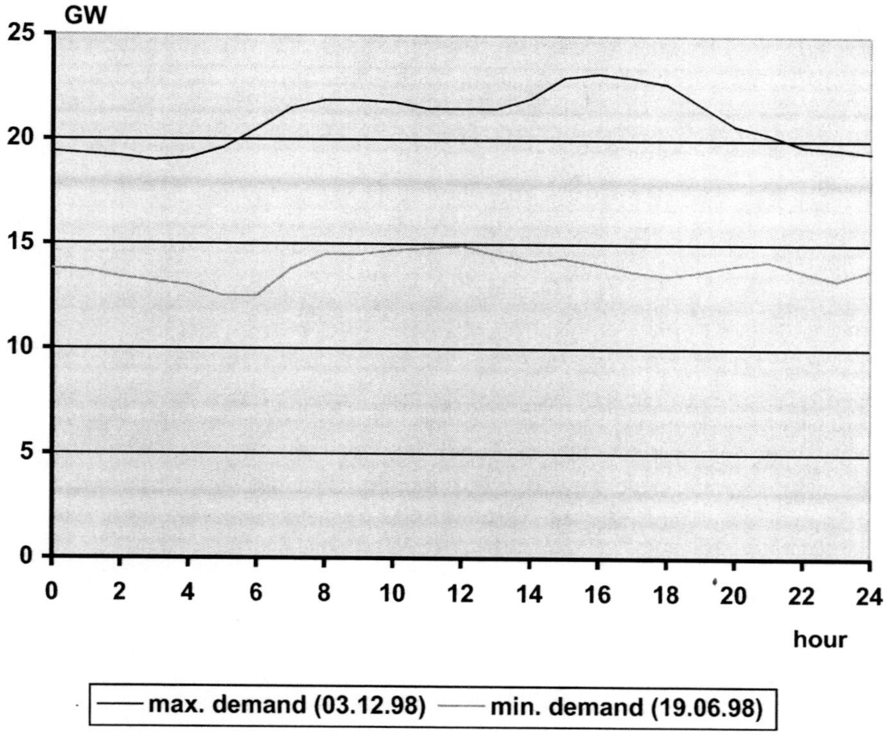

Figure 3: Maximal and minimal load curves of working days in 1998[24].

These two load curves give us an impression of the general variability of demand in Poland. From Figure 3 we can see that on Dec. 3, 1998, between 3 and 4 p.m. (15 and 16) the demand was about $D(\bar{n}, p) = 23$ GW = 23000 MW. Let us assume, that this is also the maximum demand which producers expect to be faced with – at least during this hour on this day. Assuming a linear demand function, m= 15,72 $(\text{MWh})^2$/PLN (see above), and taking into account that, in the model, p measures the difference between market price and marginal costs, we get (for 1 hour):

$$\begin{aligned} \bar{n} &= D(\bar{n}, p) + mp \\ &= 23000MWh + 15.72MWh^2 / PLN \cdot 42.74PLN / MWh \qquad (21) \\ &= 23672MWh \end{aligned}$$

42.72 PLN/MWh is the difference between the average wholesale price of 101 PLN/MWh[25] and marginal costs of 58.26 PLN/MWh.

So, the maximal demand function between 3 and 4 p.m. on Dec. 3 (and also in the whole year 1998) was

[24] Source: PSE Statystyka in: *www.pse.pl/pl/statystyka/s2*.
[25] ARE (1999 a), Tabela 12, S. 12.

$$(22)\ D(\bar{n}, p) = 23672MWh - 15.72\frac{(MWh)^2}{PLN}p,$$

where p is measured in PLN/MWh and denotes the difference between spot market price and marginal costs.

V. SIMULATION RESULTS: SPOT MARKET PRICES FOR DIFFERENT SCENARIOS

Maximal prices under the Equilibrium Selection Asssumption above are computed by inserting k, m, and n in the formulas derived in Section II. Assuming constant marginal costs, we get prices and supply (=demand) for the conditions of the hour of maximal demand in 1998 (see Table 1 and Figure 4).

Table 2: Maximal prices and supply at maximal demand ($\bar{n}$ = 23672 MWh).

Number of producers	max. Price (net of MC) [PLN/MWh] $p_{\max} = \frac{\bar{n}}{m} \cdot \frac{1}{k+1}$	Spot price P_{max}+MC [PLN/MWh]	Aggregate supply = demand. [MWh] $k \cdot S(p_{\max}) = \bar{n} \cdot \frac{k}{k+1}$
$k = 10$	136.04	194.30	21515
$k = 20$	71.68	129.94	22537

The prices p_{max} in Table 2 are net of marginal costs (= 58.26 PLN/MWh). They do not tell us anything about absolute profits, as also fix costs have to be covered, but they tell us that there are considerable differences in maximum prices under the alternative assumptions. With 10 producers price spikes of 194 PLN/MW can be expected while, with 20 producers they would be 35 % lower (130 PLN/MW). These large differences are due to the small slopes of the demand as well as the supply function. The effects are even stronger when equilibria have been adjusted to $\bar{n}$ but $n < \bar{n}$ is realised; then the price reductions compared to p_{max} or p_{max} + MC are dramatic. Let us exemplify this.

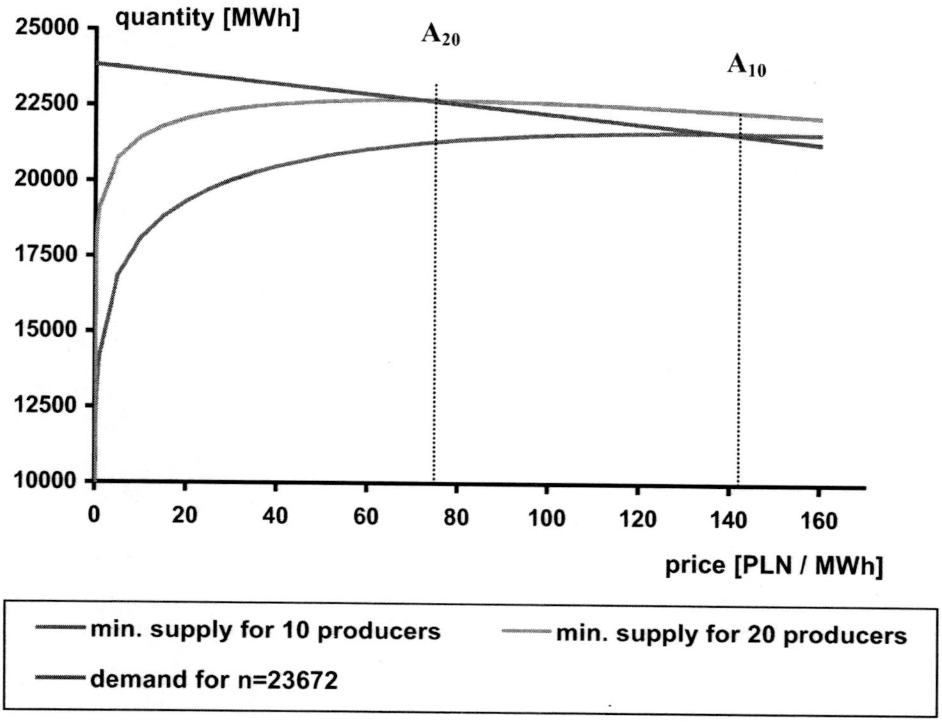

Figure 4: Maximal price and supply functions for 10 and 20 producers at maximal demand.

During the hour between 3 and 4 p.m. on Dec 3, 1998, demand was about 23 GW; during the night hours, it was about 19 GW. As, for every hour, different bids can be submitted , the producers will adjust their supply to this reduced demand by higher bids (a lower supply function). This keeps prices on a high level, namely, for 10 producers, p_{max} = 114 PLN/MW. On the weekend, when there is only about 70 % of the working day's demand (say a maximum demand of 16 GW), the price at that hour is expected to be p_{max} = 97 PLN/MW. Adding marginal costs to these values, we find that market prices are reduced by 13 % (minimum demand on Dec. 3) or 22 % (maximum weekend demand).

Results are dramatically different when producers are forced to submit bids for a longer time, for example for a day as in England and Wales under the old regime, or - why not? - even for a week: Costs and availability of power plants do not change very much during a week. Demand conditions do, however, change a lot; so demand bids should continue to be submitted hourly (or even half-hourly). What would be the consequence?

The producers would submit bids according to $\bar{n} = 23672 MWh$ (see (19)) independent of whether they apply to the hour between 3 and 4 pm on Dec. 3, or for the whole day, or even for the whole week. If the weekend demand with n = 16672 MWh, however, is matched with supply functions adjusted to $\bar{n} = 23672 MWh$ then p_{max} shrinks to 4.8 PNL/MWh, i.e. 3 % of the value at $\bar{n}$. Spot prices are reduced from 155 PLN/MWh to 63 PLN/MWh, i.e. by 60 %.

For 20 suppliers we get respective results on a lower price level.

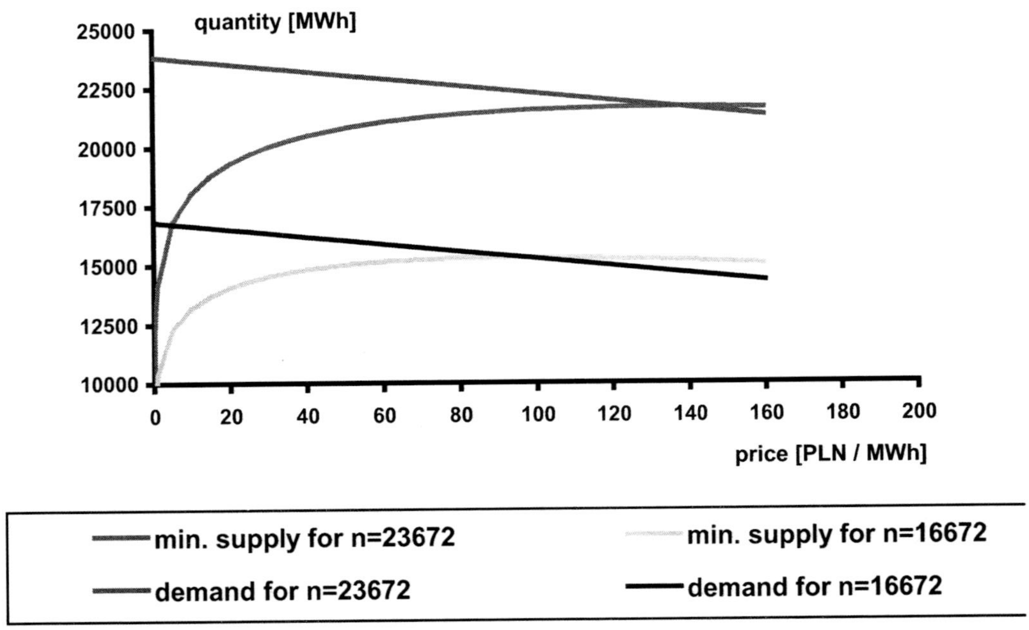

Figure 5: Adjusting or not adjusting equilibrium supply to reduced demand (linear cost function case, 10 producers).

Table 3: Prices and quantities in Figure 5.

Scenarios	**Price** p **(net of MC) [PLN / MWh]**	**p + MC [PLN]**	**Aggregate supply = demand [MWh]**
A. Minimum equilibrium supply for ($\bar{n} = 23672$).	136.04	194.30	21515
B. Supply as in A but weekend demand with $n = 16672$.	4.80	63.06	16600
C. Minimum equilibrium supply for weekend demand ($n = 16672$)	96.36	154.62	15147

If we assume a quadratic cost function as estimated in Section IV, then all computations are a bit more complicated and more tedious, but the results do not differ very much (see Kuba, 2001). Let us present here only the differences which would result in Table 3. From Table 4, we see that, for quadratic costs, prices are a bit larger which is mainly due, however, to the revised estimation of marginal costs, as we see when comparing p_{max} in Table 3 with "Spot price - MC" in Table 4.

Table 4: Adjusting or not adjusting equilibrium supply to reduced demand (quadratic cost function, 10 producers).

Scenarios	Spot price p (including MC) [PLN / MWh]	Spot price - MC [PLN/MWh]	Aggregate supply = demand [MWh]
A. Minimum equilibirum supply for maximum demand	204,84	135,92	21367,92
B. Supply as in **A** but weekend demand	69,99	6.44	16487,60
C. Minimum equilibrium demand for weekend demand	158,10	96,08	15102,72

While the consequences of introducing a quadratic cost function could not easily be foreseen, the effects of different but still linear demand functions are simpler to understand. In (14) we see that p_{max} depends on $\overline{n}/m$, so taking twice the value for m or half the value for $\overline{n}$ (because of long-term contracts) provides us with half the value of p_{max}. If p_{max} had been 100 PLN/MWh, which implies a spot price = p_{max} + MC ≈ 150 PLN/MWh, then the spot price for the reduced $\overline{n}/m$ value is about 100 PLN/MWh, i.e. it is reduced by $\frac{1}{3}$. The qualitative results from above are confirmed on a lower price level.

CONCLUSION

Do the simulated prices have a sensible order of magnitude? In Figure 6 we compare the simulated maximum prices with monthly price spikes from other market places. We left out the case of the Californian electricity crisis where wholesale prices as high on 6000 PLN/MWh were reached. But even compared with "normal" price spikes our simulation is in the region of the low price countries Norway and Switzerland. This result is not implausible as our simulation did not take into account temporary capacity shortages, which may be responsible for some of the other spikes.

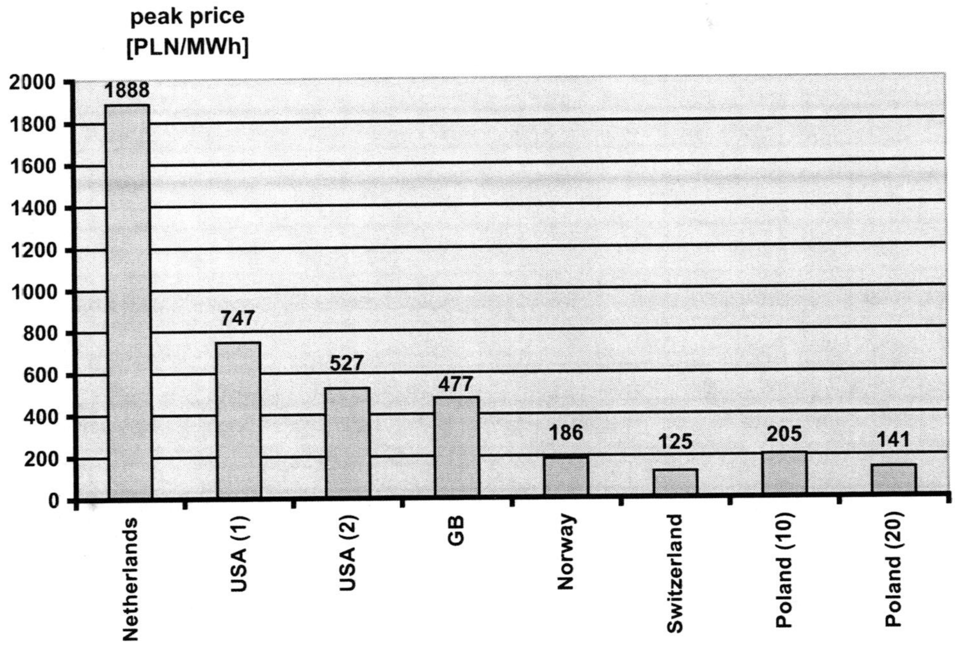

Figure 6: Peak prices in Jan. 2000 in selected countries
(Source: PIE, Feb 17, 2000) compared with our simulated prices.

The comparison shows that our results are not beyond the range of empirically found prices. Note that we did not intend to exactly forecast future Polish wholesale prices for electricity. Instead of that we wanted to show within a "realistic" model that in the electricity market

(i) even with a number of 10 to 20 producers we are far from complete competition, and
(ii) the larger the period for which producers have to submit bids (while consumers' bids are short-termed) the smaller are average spot prices; the differences can be considerable.

These results can also serve to comment on the British change of regime in 2001. The presently greater importance of the contract market has (short-term?) dampening influences on the price. The transition from daily supply bids to hourly bids, or better: the transition to a system where producers' and consumers' bids are for the same period, will (in the long run?) increase prices. Probably the greatest influence on the price development in the last years had the growing number of powerful market players. While, in 1996, PowerGen and National Power determined the spot price (by submitting the marginal bids) during 76,6 % of the periods, in 2000, six firms with shares between 10,7 % and 21,1 % were decisive altogether in 94,3 % of the periods[26]. This led to large price decreases already before 2001. The new market design was (only in the first year?) accompained by a "safety strategy" of relying more on long-term contracts. We think that the merits and the disadvantages of the new system can be evaluated only after the market participants have become acquainted to the new

[26] Source: Competition Commission, 2000.

structure. For Poland, of course, such a familiarity with (new) market structures is even more important.

The European electricity markets are still under reconstruction, even in England and Wales where there is the greatest experience with different market regimes. We can expect that Western as well as Central European countries will all re-evaluate their present structures and many will come up with new solutions. For this purpose they should not only take into account the empirical performance of extant regimes (which are often highly dependent on national circumstances) but also theoretical results as the above ones.

APPENDIX. THE INTERACTION OF FUTURES AND SPOT MARKETS

In an oligopoly of k producers without production costs and with a linerar demand function x = n – mp (an inverse demand function $p = \frac{n}{m} - \frac{1}{m}x$), the Cournot equilibrium price is $p^* = \frac{n}{m} \cdot \frac{1}{k+1}$.

If the quantity y is delivered to the consumers under firm contracts and if "efficient rationing" of this quantity[27] can be assumed, then the producers compete for a "rest demand" x – y = n – y – mp which results in an equilibrium price $p^* = \frac{n-y}{m} \cdot \frac{1}{k+1}$. So, on first glance, the price consumers have to pay decreases considerably because of the long term contracts they have concluded.

This is still true on the second glance, namely when we take into account also the market for futures or long term contracts (see Allaz, 1992; Powell, 1993; Bolle, 1993); but whether the average price is "considerably" reduced is questionable. The strategically acting producers know that the price q which can be obtained on the contract market depends on the expected spot market price p: The lower p, the lower the demand on the contract market for the same q. Thus, in a market with experienced participants the spot market will have a considerable volume. The difference between q and p is due to the conditions (stochastic elements) of the spot market, the risk aversion of the participants, and the degree of concentration of producers.

The more risk averse the consumers are, the higher is q compared with p. (If there are non-strategically acting risk neutral participants in the market, the we get p = q.) So, if we model only a spot market instead of a hierarchy of contract and spot markets we should perhaps take the result as a proxy for average contract and spot prices. The real average is to be expected to be smaller than the result of the spot market simulation but it is difficult to say how much smaller it is.

[27] See Tirole, 1992.

REFERENCES

Allaz, B. (1992), "Oligopoly, uncertainty and strategic forward transactions", *International Journal of Industrial Organization, 297-308.*

ARE (1999 a): "Informacja statystyczna o energii elektrycznej – Biuletyn miesięczny", *Agencja Rynku Energii, 6 (66), 06.1999.*

ARE (1999 b): "Ubytki mocy w elektrowniach zawodowych 1998", *Agencja Rynku Energii, Warszawa, 07.1999.*

Ausubel, L. M. and Crampton, P. C. (1998): "Demand Reduction and Inefficiency in Multi-Unit Auction", *Working Paper, University of Maryland.*

Bolle, F. (1989): "Deregulated Spot markets for Electricity: a Continuum of Solutions and the Danger of Tacit Collusion", *presented at the meeting of the Econometric Society, München, Sept. 1989.*

Bolle, F. (1989): "Supply function equilibria and the danger of tacit collusion – The case of spot markets for electricity", *Energy Economics, 4.1992, S. 94-102.*

Bolle, F. (1993): "Who profits from Futures Markets", *ifo Studien 39, 239-256.*

Bolle, F. (1997): "Necessary Conditions for Efficient Multiple-Bid Auctions" *in Economic and Environmental Risk and Uncertainty: New Models and Methods (edited by R. Nau, E. Gronn, M. Machina and O. Bergland), 123-143, Kluwer Academic Publisher.*

Bolle, F. (2001): "Competition with supply and demand functions", *Energy Economics, 23 (2001), S. 253-277.*

Bolle, F. and Breitmoser, Y (2002): "Spectrum Auctions: How They Should and How They Should Not Be Shaped, *to appear in Finanzarchiv.*

Competition Commission (2000): "AES and British Energy. A report on references made under section 12 of the Electricity Act 1989"

Dowzycki, A. (2002): "KONKURENCJA I WSPÓLPRACA W PRZEMYSLE ELEKTROENERGETYCZNYM PO OBU STRONACH ODRY (ze szczególnym uwzglednieniem podsektora dystrybucji energii elektrycznej)", *Manuscript, Poznan.*

Engelbrecht-Wiggans, R. (1988): "Revenue Equivalence in Multi-Object Auctions", *Economic Letters 26, 15-19.*

Fehr, N.-H. M. von der, Harbord, D. (1993): "An Auction Approach to the Study of Spot-Market Competition in a Deregulated Electricity Industry", *Economic Journal 103, 531-546.*

Green, R. J. and Newbery, D. M. (1992): "Competition in the British electricity spot market", *in Journal of Political Economy, 100 (1992), S. 929-953.*

Green, R. J. (1996): "Increasing Competition in the British Electricity Spot Market", *in Journal of Industrial Economics, Vol. 44, No. 2, Blackwell Publishers Ltd., 06/1996, S. 205-216.*

Klemperer, P.D. and Meyer, M. (1989): "Supply function equilibria in oligopoly under uncertainty*", in: Econometrica, Vol. 57, S. 1243-1277.*

Kuba, E. (1998): "Die Umstrukturierung der polnischen Elektrizitätswirtschaft", *Zeitschrift für Energiewirtschaft, 2/98, S. 110-122.*

Kuba, E. (2001): "Die Umstrukturierung der polnischen Elektrizitätswirtschaft im Lichte der Liberalisierung in Großbritannien und Norwegen - Vergleich und theoretische Analyse", Europa-Universität Viadrina, Dissertation, eingereicht im Oktober 2001.

Nautz, D. (1994): "Bidding Behavior in the Zinstender of the German Bundesbank: Theory and Evidence", *Manuscript, Freie Universität Berlin.*

Powell, A (1993): Trading Forward in an Imperfect Market. The Case of Electricity in Great Britain", *The Economic Journal 103, 444-453.*

Tirole, J. (1992): "The Theory of Industrial Organization", *MIT Press.*

Wolfram, C. D. (1998): "Strategic bidding as a multiunit auction: on the emperical analysis of bids to supply electricity in England and Wales", *RAND Journal of Economics 29, 4, 703-725.*

Chapter 2

CARBON EMISSIONS CONTROL, TRADE LIBERALIZATION AND ELECTRICITY MARKET IMPERFECTION: COORDINATED APPROACHES TO TAIWAN'S TRADE AND TAX POLICY

Hao-Yen Yang
Associate Professor
Department of Economics
Shih-Hsin University
No.1, Lane 17, Mu-Cha Rd. Sec. 1, Taipei, Taiwan, 116
Phone: +886-2-22368225 ext.3408
Fax: +886-2-22361658
e-mail: haoyen@cc.shu.edu.tw

ABSTRACT

A neoclassical and static computable general equilibrium model has been applied to investigate the linkages between trade and environmental policies. The objective of this paper is to offer a quantitative analysis of the joint effects of trade and environmental policies on the Taiwan economy. My estimates are derived from five-household, 18-sector computable general equilibrium model calibrated to a 1999 Taiwan's social accounting matrix. The empirical results show that, under a perfectly competitive electricity market, tariff elimination can significantly reduce the costs of carbon emissions control and possibly offset welfare reductions due to implementation of a carbon tax. This finding reveals that the coordination of trade liberalization with the introduction of targeted environmental policies enhances factor relocation towards competitive industries; they are broadly consistent with those cases of developing countries reported in previous studies. However, when the electricity market becomes less competition as a result, the joint effect of these two policies on economic efficiency is insignificant. This paper shows that jointing trade and environmental policies cannot attain the efficiency objectives under assumption of non-competitive electricity market. This result suggests that the conditions of market structure play an important role for Taiwan implementing trade and environmental policies; and such conditions affect the resource pulls in the Taiwanese economy, thereby influencing the magnitude of the efficiency change from these two policies.

Keywords: Carbon emissions control, trade liberalization, electricity market imperfection, computable general equilibrium, NICs, Taiwan

1. INTRODUCTION

Taiwan is experimenting with trade liberalization, while being aware of environmental problems, and the coherence between trade and environmental policies is a major concern for design of development strategies.

Recently, several studies have emerged on the possible reconciliation between trade and environmental policies. Important linkages have been identified, and major conclusions have been drawn in terms of economic policy (see for instance Perroni and Wigle, 1994; Lee and Roland-Holst, 1997; Beghin *et al.*, 1995; Dessus and Bussolo, 1998). The coordination of trade liberalization with the introduction of targeted environmental policies can produce economic gains, partly because there exists a real comparative advantage in dirty productions. Most empirical studies tend to confirm that developing economy specialized in dirty industries, and hence a trade-off between trade liberalization and environmental preservation could occur (see Beghin *et al.*, 1994; Dessus and Bussolo, 1998, for surveys). Although previous studies have involved a detailed evaluation of the linkages between trade and environmental policies, little attention has been devoted to inquiring about the economic implications of the linkages of these policies in the newly industrialized countries (NICs). My concern here is to investigate this issue in Taiwan.

The existing applied modeling studies the relationships between trade and environmental policies have generally relied on computable general equilibrium model with perfectly competitive specifications. However, it is the perception of this paper that non-competitive market structures and regulated managerial practices are pervasive facts in many developing and industrialized economies. At a theoretical level, the implications of environmental policy in monopolized product markets have been studies in a series of seminal papers (see Roe and Yeldan, 1996, for surveys). These results have shown that under a second best setting of non-competitive, oligopolistic market composition, theoretically sound environmental policies may not attain the desired objectives. In addition, many authors have observed that pro-competitive forces operating on imperfectly competitive firms will play an important role in countries undergoing trade liberalization (see for instance Devarajan and Rodrik, 1991). Such forces affect the resource pulls in the economy, thereby influencing the magnitude of the efficiency gains from liberalization. Therefore, it is argued in this paper that, to the extent of market imperfection exists in the Taiwan's economy, trade and environmental policies may generate different results in this objectives.

The objective of this paper is to investigate the interactive effects of trade and environmental policies on the Taiwanese economy. The alternative that we explore in this paper continues to assume a World Bank type of computable general equilibrium model, but drops the assumption that electric power sector produces for a perfectly competitive market. Indeed, I consider the effects of modeling of imperfect competition in the electricity market, introduced in my previous papers (Yang, 2001a, 2001b). Now the electric power sector in Taiwan exhibits just those features that have been the focus of literature. At present Taiwan has only one integrated power company-the state-owned Taiwan power Co (Tai-power)-which not only produces power but also distributes it. Like many other NICs, Taiwan is experiencing a rapid trade-led growth, with a sustained high rate of growth of demand for electricity. Until now, the energy-related carbon emissions from the electric power sector are approximately one-third of total carbon dioxide emissions (Yang, 2001a). What is the interactive effect of trade and environmental policies in Taiwan? Will trade liberalization reduce the costs of carbon emission controls and possibly eliminate aggregate welfare and output reductions due to the implementation of a carbon tax in an oligopolistic electricity market environment versus a perfectly competitive setting? This paper aims at offering a quantitative analysis of the linkages between economic activity and the environment in Taiwan, and specifically at evaluating the joint effects of tariff and carbon taxes.

This paper first considers interactive effects of tariff and carbon taxes under a perfectly competitive electricity market, and examines its effects on gross domestic product, resource allocation and income distribution. In a second scenario, I show how free trade can't mitigate the anti-growth effects of environmental tax policy under an imperfectly competitive electricity market with mark-up pricing. The organization of the paper is as follows. After this introduction, in the next section I briefly describe the characteristics of the Taiwanese economy. Thereafter, I introduce a computable general equilibrium model and present the data sources. Then, the simulation results will be discussed. The final section gives our summary and conclusions. Appendix A is a presentation of Taiwan's computable general equilibrium model in a complete equation form.

2. The Taiwanese Economy

Taiwan is an island with a small area of about 36,000 square kilometer. Although Taiwan is not endowed with rich natural resources on land, this island has achieved an amazingly high economic growth rate and has been very successful in pursuit of its economic development goals.

Like many other NICs, Taiwan has followed a trade-led growth strategy since the early 1960s. The ratio of export with respect to gross national product (GNP) increased form 11.51% in 1960 to 53.66% in 2000, while that of import rose from 19.03 to 51.39%. The high ratio of export/GNP and import/GNP both averaging 50 percent make its economy very sensitive to the trade policy (see Table 1). Disillusioned with the import-substitution policies of the past, Taiwan is experimenting with trade liberalization.

Table 1. Primary economic condition of Taiwan

Year	GNP (NT$ million, in 1991 prices)	Total energy consumption (1000KLOE)	$\frac{export}{GNP}$	$\frac{import}{GNP}$
1960	327,817	3,839	11.51	19.03
1965	525,097	5,779	19.36	22.36
1970	822,457	9,870	30.36	30.41
1975	1,255,539	15,915	39.85	43.15
1980	2,084,609	29,579	52.60	53.79
1985	2,936,389	36,233	53.32	39.75
1990	4,581,783	52,007	45.64	40.78
1995	6,400,966	68,964	48.26	46.31
2000	8,439,804	90,911	53.66	51.39

Source: AREMOS economic-statistic databanks.

In 1960, Taiwan's real GNP was NT$327,817 million while that of 2000 was NT$8,439,804 million (Table 1). The real economic growth rate of GNP has reached an average of 8.30 per year during the past 41 years (1960-2000). The rapid economic growth and development has created substantial changes in production ratios for the nation's economic sectors. The size of the agricultural sector as a percentage of gross domestic product dropped from 27.45% in 1960 to 2.09% in 2000; industrial production increased from 26.57% to 32.38%, while that of services rose from 45.98% to 65.53% (Table 2). Before 1970, Taiwan's industry was labor-intensive emphasizing products such as textiles, apparel, electronic components. Heavy industries, such as petrochemicals, steel, electrical equipment, and automobile industry were subsequently introduced leading to a more energy-intensive economy (Hsu, Leung, and Ching, 1987). The structural change in Taiwan economy has created a rapid increase in energy consumption. Energy consumption increased from 3,839 thousand kiloliters of oil equivalent in 1960 to 90,911 thousand kiloliters in 2000, at average annual growth rate of 8.50% (Table 1). Thus, the long-run energy elasticity with respect to GNP is slightly above 1.0.

Table 2. Structure of Taiwanese economy

Year	Sectoral distribution of gross domestic product (%)		
	Agriculture	Industry	Services
1960	27.45	26.57	45.98
1965	23.63	30.21	46.16
1970	15.47	36.83	47.70
1975	12.70	39.92	47.38
1980	7.68	45.75	46.57
1985	5.78	46.28	47.94
1990	4.18	41.23	54.59
1995	3.48	36.37	60.15
2000	2.09	32.38	65.53

Source: AREMOS economic-statistic databanks.

The energy demand for Taiwan's high growth economy soared over the past several decades. Since Taiwan is not endowed with rich energy resources on land, the ratio of indigenous energy to total energy supply decreased from 69.47% in 1962 to 2.86% in 2000 while that of imported energy increased from 30.53% in 1962 to 97.14% in 2000 (Table 3). In order to copy with the actual energy demand and coordinate with the domestic economic development, the government of Taiwan must assure the stability of energy supply. Before 1966, coal was the main energy source in Taiwan, but oil has replaced coal as a major energy source since 1967. Since the early 1970s (the second oil crisis), the government of Taiwan has advocated the substitution of nuclear energy for oil. The first nuclear power plant was merged into the system in 1977. At present coal, oil, natural gas, hydro, nuclear resources are used in combination (see Table 3).

Table 3. Structure domestic composition of energy supply

Year	By indigenous or imported (%)		By energy form (%)					
	Indigenous	Imported	Coal	Petroleum	Liquefied natural gas	Natural gas	Hydro power	Nuclear power
1962	69.47	30.53	58.66	30.57	-	0.71	10.06	-
1965	66.16	33.84	51.70	34.15	-	4.59	9.56	-
1970	40.71	59.29	26.88	59.18	-	7.96	5.98	-
1975	30.80	69.20	13.13	70.00	-	9.22	7.65	-
1980	13.62	86.38	15.21	71.02	-	5.71	2.12	5.94
1985	11.25	88.75	21.74	52.49	-	3.36	4.35	18.06
1990	6.56	93.44	23.35	55.42	1.61	2.23	3.47	13.39
1995	4.22	95.78	26.19	54.28	4.58	1.17	2.79	11.01
2000	2.86	97.14	30.78	51.44	6.01	0.70	2.07	9.00

Source: AREMOS economic-statistic databanks.

After the two energy crises in 1970s, the energy policy for the Taiwan area has been made in accordance with the characteristics of the energy supply and demand in Taiwan and global energy trends. However, owing to the energy environmental changes at home and abroad in the recent years and such problems as the greenhouse effects resulting from the utilization of fossil fuels that have aroused much international concern. Therefore, the coherence between trade and environmental policies is a major concern for design of energy strategies. Hence, the question I ask in this paper, 'What are the consequences of trade and environmental policies for an economy that exhibits the characteristics of Taiwan?' is of interest to the country itself.

3. Model Specification

The computable general equilibrium model I used is a static model which follows closely in the tradition of Dervis, de Melo and Robinson (1982). My modifications include various electricity market structures, alternative function forms for production technology and disaggregate household groups. In my computable general equilibrium model, Taiwan's economy is divided into eighteen sectors: (1) agriculture (i.e. agriculture, forestry and fishing), (2) coal mining, (3) crude oil, (4) natural gas, (5) other mining, (6) food and tobacco, (7) textiles and apparels, (8) wood and bamboo, (9) chemicals and plastics, (10) petroleum products, (11) coal products, (12) non-metallic mineral products, (13) metal products, (14) electricity, (15) gas supply, (16) other manufacturing, (17) transport and (18) finance, service and others. We also specify the five equal divisions of household, using the classification of *The Report on the Survey of Family Income and Expenditure in Taiwan area* (DGBAS, 1999b). A disaggregate social accounting matrix given in table 4 is designed to reflect the circular flow of the Taiwanese economy that is embodied in Taiwan's computable general equilibrium model which we will outline below. The model is composed of production and factor markets, expenditure and income, foreign exchange market, prices, market equilibrium and macro close rule. The following paragraphs provide a general description of Taiwan's computable general equilibrium model. Appendix presents the complete set of equations describing the model.

The production technology is specified that each economic sector combines capital, labor, electricity, fossil fuel and other intermediate, using nested constant elasticity of substitution (CES)-Leontief production function designed to product a specific commodity of sectors. The formulation separates fossil fuels and electricity from other intermediate goods and from capital and labor. This enables us to allow for easier substitution between fuel inputs than between fuel and non-fuel inputs. Coal is the most carbon-intensive fuel per unit of energy and natural gas the least. Thus, aggregate carbon dioxide emissions are given by:

$$e = \varphi_{coal} X_{coal} + \varphi_{oil} X_{oil} + \varphi_{gas} X_{gas} \quad (1)$$

where φ_f is carbon emissions per unit of fossil fuel f , X_f is products of fossil fuel f . This specification assumes that emissions are proportional to fuel use - there are currently no economically viable end-of-pipe abatement technologies for scrubbing emissions before they enter atmosphere (see Parry and Williams, 1998).

Both input and output markets (excluding electricity market) operate under perfectly competitive market conditions, and specify the behavior of optimizing producers in the market economy. Factor demands are derived from first-order conditions, using a conventional profit maximization approach. Labor and capital factors are perfectly mobile between economic sectors and are fully employed in the base year.

In my electricity market specifications, it is a widely accepted fact that public enterprise pricing is more often a politically regulated process, and the behavior of mark-ups can be regarded to be mostly under the influence of the private directions (Roe and Yeldan, 1996). Therefore, I assume that electricity market operates under imperfectly competitive market conditions, and determine commodity price according to the mark-up pricing rule:

$$P_i = (1+m_i)\left(\frac{\lambda_i^w wL_i + \sum_j P_j a_{ji} X_i}{X_i(1-td_i) - P_{CO_2} TQ_{CO_2,i}}\right); i \in \textit{electricity market} \quad (2)$$

where P_i is electricity price, m_i is the mark-up rate, λ_i^w is labor market distortion parameter, w is aggregate labor factor price, L_i is labor input, P_j is other goods price, a_{ji} is input-output coefficient, X_i is production output, td_i is indirect tax rate of domestic goods, P_{CO_2} is carbon price, and $TQ_{CO_2,i}$ is total carbon emissions. This specification assumes electricity price is set as average variable costs plus a mark-up, and with price being pre-determined as such, electricity output is driven by aggregate demand.

My model has five household groups, for each household group there is a representative consumer who determines his consumption behavior from utility maximization. The consumer preferences are characterized by Cobb-Douglas utility function for each representative consumer, this allows for different marginal budget shares for five household groups. The household budget account constitutes income revenue and expenditure. The main sources of household income revenue include labor income, distributed profits, government transfer and net transfer from abroad. Each household's expenditure is allocated to private consumption, direct taxes and private saving.

Government actions have a significant influence on the economy through its income revenue and spending changes. The main sources of government revenues include indirect taxes, tariffs, enterprise taxes and direct taxes. Government spending constitutes government consumption, transfer to enterprise, transfer to household and government saving. Our model assumes government income revenue equals to income spending.

The enterprise in this model includes private companies, public companies and non-profit enterprise, using the aggregate of catalog of *National Income in Taiwan area* (DGBAS, 1999c). The main sources of enterprise revenues include capital income, government transfer and net transfer from abroad. Main enterprise outlays include distributed profits, enterprise taxes and retained earnings. We also assume enterprise income revenues equals to outlays.

In my model, small-open-economy assumption implies Taiwan cannot affect world price for its exports and imports. The model assumes float exchange rate mechanism and contains a balance-of-trade constraint. The value of total exports includes export trade, net labor income from abroad, net capital income from abroad, net transfer from abroad to enterprise and net transfer from abroad to household. The value of total imports consists of import trade and net foreign borrowing.

Table 4. Structure of a social accounting matrix for Taiwan, 1999 (Million NT)

	Expenditures								
			Factors		Institutions			Capital	
Receipts	Activities	Commodities	Labor	Capital	Enterprises	Households	Government	Account	Rest of World
Activities		Domestic sales 14,633,364							Exports 4,402,101
Commodities	Intermediate inputs 9,884,884					Private consumption 5,641,313	Government consumption 1,221,717	Investment 2,255,365	
Factors:									
Labor	Wages 4,687,235								Net labor income from abroad -17,032
Capital	Rentals 3,106,004								Net capital income from abroad 102,944
Institutions:									
Enterprises				Capital income 3,208,948			Government transfers -56,174		
Households			Labor income 4,670,203		Distributed profits 2,393,553		Government transfers 346,323		Net transfer from abroad -71,860
Government	Indirect taxes 553,198	Tariffs 139,348			Enterprise taxes 316,707	Direct taxes 526,765			Net transfer from abroad 1,460
Capital Account	Capital depreciation 804,144				Retained earnings 442,514	Private saving 1,170,141	Government saving 25,612		

	Expenditures								
			Factors		Institutions			Capital	
Receipts	Activities	Commodities	Labor	Capital	Enterprises	Households	Government	Account	Rest of World
Rest of world		Imports 4,230,567						Net foreign borrowing 187,046	
Totals	Total costs 19,035,465	Total absorption 19,003,279	Labor income 4,670,203	Capital income 3,208,948	Enterprise income 3,152,774	Household income 7,338,219	Government income 1,537,478	Total investment 2,442,411	Foreign exchange inflow 4,417,613

Source: DGBAS (1999b)

The model assumes imperfect substitution between imports and domestic goods, using the Armington constant elasticity of substitution function form. Thus, the import demand for a specific sector becomes derive demand. On the export side, the export demand for a specific commodity depend on the world price of exported goods measured in foreign exchange units relative to price level of domestic goods.

In a market equilibrium condition, I assume that the total supply of commodities equals the total demand for the commodities. As stated previously, total supply of composite commodities is a CES combination of domestic goods and imports. Total demand consists of intermediate inputs, private consumption, government consumption and investment.

Finally, because the model is closed in that it satisfies Walras' Law, the macro closure condition requires that aggregate saving equals aggregate investment. Aggregate investment is savings-driven, this specification is called neoclassical closure in the general equilibrium literature. In computable general equilibrium model, the system can only determine relative prices, and solve for prices relative to a numeraire. In our model the numeraire is the gross domestic product price deflator.

To implement the model, I have selected 1999 as the benchmark year as the best recent, comprehensive and consistent data was available in the form of a social account matrix. A disaggregate social accounting matrix combines the *input-output table* (DGBAS, 1999a) with the national income accounting. Therefore, I base disaggregate social accounting matrix on the national income accounting, and adjust input-output data for consistency. In addition to the data in the input-output table and national income accounting, the database for my computable general equilibrium model includes labor, capital, income distribution matrix, various elasticities, other parameters and carbon emissions. The labor and capital data are from *Year Book of Manpower Statistics* (DGBAS, 1999d) and *The Trends in Multifactor Productivity* (DGBAS, 1999e). A multi-sector income distribution matrix is transformed income flows between industry and the five equal divisions of household (see Hanson and Rose, 1997). This matrix used is from *The Report of Survey of Family Income and Expenditure* (DGBAS, 1999b). In order to model production technology, composite commodity and export demand function, I need a variety of elasticity. Elasticities of production technology are constructed on the basis of the elasticities used in Burniaux et al. (1992). Following Chu (1996) I set the elasticity between domestic and imported goods and elasticity of export demand function. Table 5 reports the various elasticities used in the simulations. Other parameters in this model are numerically specified by calibrating against a benchmark equilibrium data set which is derived from above. The data on carbon emissions used in this study are compiled from Huang, Yang and Chu (1993) and the AREMOS economic-statistic databanks. Simulation results are generated by using GAMS program (Brook, Kendrick, and Meeraus, 1992).

Table 5. Various elasticity specifications

Sector	Elasticity of Export Demand	Elasticity of Domestic and Imports
1 Agriculture	0.65	4.00
2 Coal mining	1.45	1.50
3 Crude oil	1.45	1.50
4 Natural gas	1.45	1.50
5 Other mining	1.45	1.50
6 Food and tobacco	0.75	3.57
7 Textiles and apparels	1.05	2.58
8 Wood and bamboo	1.05	2.58
9 Chemicals and plastics	1.05	3.61
10 Petroleum products	1.00	2.30
11 Coal products	1.00	2.30
12 Non-metallic mineral products	1.00	2.30
13 Metal products	1.50	1.53
14 Electricity	0.00	0.33
15 Gas supply	0.00	0.33
16 Other manufacturing	1.15	2.33
17 Transport	0.00	0.90
18 Finance, service and others	0.00	0.90

Notes: 1. Ranges of Elasticity of substitution between capital and labor are 0.5-0.7.
2. Ranges of Elasticity of substitution between electricity and non-electricity energy are 0.9-1.2.
3. Ranges of Elasticity of substitution between capital-labor combination and energy are 0.20-0.30.

4. EMPIRICAL RESULTS

This paper is to analyze the joint effects of tariff and carbon taxes under a perfectly competitive electricity market (scenario I) versus a non-competitive environment (scenario II). I simulate a 1% carbon emission reduction via a carbon tax, and the tariff rate reduces to the level which keeps government deficit unchanged in these two scenarios. Table 6 provides the main aggregate indicators from the simulations: CO_2 emissions, tax per ton CO_2, gross domestic product, value-added, private consumption, equivalent variations, factor relocation and income distribution indices.

Table 6. Simulation results

Variable	Simulations	
	I	II
CO_2 emissions	-1.000	-1.000
Tax per ton CO_2 (NT)	161.3750	201.1192
Gross domestic product	0.0215	0.0317
Value-added	0.0143	-0.0117
Private Consumption	-0.0135	-0.0213
Equivalent Variation (EV)		
Income Class 1 (lowest)	-0.0774	-0.0813
Income Class 2	-0.0501	-0.0543
Income Class 3	-0.0469	-0.0501
Income Class 4	-0.0597	-0.0536
Income Class 5 (highest)	-0.0673	-0.0712
Factor relocation		
Labor	0.001017	0.001119
Capital	0.001691	0.001753
Gini coefficient	0.0021	0.0025
Disposable Income		
Income Class 1 (lowest)	-0.0381	-0.0417
Income Class 2	-0.0061	-0.0093
Income Class 3	-0.0013	-0.0018
Income Class 4	-0.0111	-0.0132
Income Class 5 (highest)	-0.0135	-0.0235

Notes: 1. Value-added does not include import taxes and carbon taxes.

2. Values of gross domestic product, value-added, private consumption, Gini coefficient, and disposable income are percentage change from baseline. Equivalent variations base on Laspeyres' index with respect to private consumption.

3. Factor relocation is the ratio of factor mobility with respect to total factor supply.

4.Tax per ton CO_2 is in 1999 dollars.

I now investigate the joint effects of tariff and carbon taxes under a perfectly competitive electricity market. With respect to gross domestic product effects, my results show a 0.0215% gross domestic product gains under scenario I (Table 6). It also shows the value-added effect is positive. According to my model specifications, meeting carbon emissions targets via a carbon tax increases the price of carbon-embodied goods. As price rises, production output changes among sectors due to factor mobility and inter-industry effects. At a theoretical view, the overall economic effect will decrease in the value-added at constant price due to a change in the price of carbon-embodied goods. However, a decrease in tariff will cause the economy to use its resources more efficiently among sectors. In this scenario, the reduction of tariff generates higher efficiency improvement, so that the joint effect procures higher net gross domestic product and a rise in the value-added. In general, the value-added effect is often recommended as a static resource allocation effect. Thus, a value-added increase indicates that the economy uses its resources more efficiently among sectors.

Table 6 (scenario I) shows that joint effects of tariff and carbon taxes under a perfectly competitive electricity market leads to a 0.0135% decrease in the private consumption. In order to discuss the consumer's welfare effect of the implementation of these two policies, equivalent variation is employed which is widely used in computable general equilibrium studies. In scenario I, the equivalent variation is the maximum amount the consumers would be presented to pay at the budget level of scenario I to avoid the increase in prices from the baseline to scenario I. That is, a welfare-improving change means that the equivalent variation is positive, while a welfare-worsening change means that the equivalent variation is negative. In this study, equivalent variation based on the Laspeyres' index with respect to private consumption is used. Since my model has five household groups, for each household group there is a representative consumer who determines his consumption behavior from utility maximization. Therefore, Table 6 (scenario I) shows five equivalent variations for different household groups. Income class 1 is the lowest income group, while income class 5 is the highest income group. As shown in Table 6, the coordination of tariff and carbon tax policies causes welfare loss in five household groups. Although the reduction of tariff increases welfare and economic efficiency, such effect of tariff decreasing cannot offset the welfare reductions due to implementation of a carbon tax in scenario I. Moreover, the lowest income group has relatively large welfare loss. This is partly because the lowest income group spends a larger part of its income on carbon-embodied goods. Another reason is that a rise in the price of carbon-embodied goods and a fall in the disposable income reduce their consumption ability. However, the highest income group does not have relatively small welfare loss. Our results reveal that the coordination of tariff and carbon taxes policies appears to be mildly regressive. This finding seems interesting for social and political acceptability in trade and environmental policies.

Table 6 (scenario I) shows that joint effects of these two policies causes overall household's disposable income to decrease. The distributional effects are summarized by the changes in the Gini coefficient. As shown in Table 6 (scenario I), there is a worsening in the distribution of income reflected by a slight rise in the Gini coefficient. As a result, the lowest income group has relatively large disposable income loss, while the highest income group does not have relatively small income decrease. This is partly because carbon taxes cause the price of capital to fall relative to labor. The highest income group derives a larger fraction of the income from capital ownership. Thus, the highest income group is hurt to a relatively larger degree on the income side.

Next, we investigate scenario II where carbon taxes and tariff elimination are jointly implemented under an imperfectly competitive electricity market with mark-up pricing. With respect to value-added effects, my results show a 0.0117% value-added loss under scenario II. This result implies that the coordination of trade liberalization with the introduction of targeted environmental policies under imperfect environment enhances factor relocation towards less efficient industries. Table 6 shows that joint effects of tariff and carbon tax policies in scenario II has a higher ratio of factor relocation by comparing the implementation of these two policies with scenario I. Table 6 (scenario II) also shows that the coordination of these two policies leads to a 0.0213% decrease in the private consumption, and causes overall welfare loss and household's disposable income decreases. Although the net welfare change from joint effects of these two policies is also negative, the tariff elimination offsets the welfare cost of the carbon taxes to a much littler extent under Scenario I.

The question naturally arises about how sensitive the model results are to changes in the values of the various parameters. To test the sensitivity of some important model results to these parameters, a method used by de Melo *et al.* (1989) was adopted. This method involves computing &low (high) elasticities for the key parameters. In most simulations, I rely on the above values for elasticities which I refer to as the central elasticity case. The low (high) elasticity case is obtained by reducing (augmenting) the values of the elasticities by one standard deviation. Results from these experiments show that one standard deviation changes in the elasticity have little effect on the model results associated with the joint effects of these policies. Thus, if the sensitivity of the results can be taken as a general guide, they suggest that the results of the experiments undertaken in this study are relatively robust around the current values of the key parameters.

5. CONCLUSIONS

A neoclassical and static computable general equilibrium model has been applied to investigate the linkages between trade and environmental policies. I extend the Dervis *et al.* (1982) model to incorporate various electricity market structures, alternative function forms for production technology, disaggregate household groups and carbon emission equation. The purpose of this paper is to offer a quantitative analysis of the joint effects of trade and environmental policies on the Taiwan economy.

My empirical results based on Taiwan's data show that, under a perfectly competitive electricity market, tariff elimination can significantly reduce the costs of carbon emissions control and possibly offset welfare reductions due to implementation of a carbon tax. This finding reveals that the coordination of trade liberalization with the introduction of targeted environmental policies enhances factor relocation towards competitive industries; they are broadly consistent with those cases of developing countries reported in previous studies (see for instance Lee and Roland-Holst, 1997; Dessus and Bussolo, 1998). However, when the electricity market becomes less competition as a result, the joint effect of these two policies on economic efficiency is insignificant. This paper shows that jointing trade and environmental policies cannot attain the efficiency objectives under assumption of non-competitive electricity market. This result suggests that the conditions of market structure play an important role for Taiwan implementing trade and environmental policies; and such conditions affect the resource pulls in the Taiwanese economy, thereby influencing the magnitude of the efficiency change from these two policies.

I acknowledge that my analysis has involved various simplifying assumptions, and that these may call my results into question. For example, I only evaluate the economic cost implications of the carbon tax. The benefits of the carbon tax and the introduction of cleaner technologies are not analyzed in this paper. In addition, the model does not consider other emission taxes. Further investigation of various emission taxes would provide better information about the economic consequences of linkages between trade and environmental policies. I also realize that other features of the model, balance trade and other distortions in taxation, are likely to dampen the impact of both trade and environmental policies. Therefore, my simulation results are only the first step of policy making. Future analyses are needed to explore the influence of these conditions on the linkages between trade and environmental policies.

REFERENCES

Beghin, J., Roland-Holst, D., van der Mensbrugghe, D., 1994. A survey of the trade and environment nexus: global dimensions, *OECD Economic Studies* 23, 167-192.

Beghin, J., Roland-Holst, D., van der Mensbrugghe, D., 1995. Trade liberalization and the environment in the Pacific basin: coordinated approaches to Mexican trade and environment policy, *American Journal of Agricultural Economics* 77, 778-785.

Brook, A., Kendrick, D., Meeraus, A., 1992. *GAMS: A User's Guide*, The Scientific Press, South San Francisco.

Burniaux, J.M., Martin, J.P., Nicoletti, G., Martins, J.O., 1992. GREEN: a multi-sector, multi-region dynamic general equilibrium model for quantifying the costs of cubing CO2 emissions: a technical manual, *Economics Department Working Papers*, No. 116, OECD, Paris.

Chu, Y.P., 1996. Effects of trade liberalization on Taiwan's economy: a computable general equilibrium model, *Working Paper*, Academia Sinica, Taipei.

De Melo, J., Stanton, J., Tarr, D., 1989. Revenue-rising taxes: general equilibrium evaluation of alternative taxation in U.S. petroleum industries, *Journal of Policy Modeling* 11(3), 425-449.

Dervis, K., De Melo, J., Robinson, S., 1982. *General Equilibrium Models for Development Policy*, Cambridge University Press, Cambridge, MA.

Dessus, S., Bussolo, M., 1998. Is there a trade-off between trade liberalization and pollution abatement: a computable general equilibrium assessment applied to Costa Rica, *Journal of Policy Modeling* 20, 11-31.

Devarajan, S., Rodrik, D., 1991. Pro-competitive effects of trade reform: Results from a CGE model of Cameroon, *European Economic Review* 35, 1157-1184.

Directorate-General of Budget, Accounting and Statistics (DGBAS), 1999a. *Input-output table*, Executive Yuan, Taiwan, ROC.

Directorate-General of Budget, Accounting and Statistics (DGBAS), 1999b. *The report on the survey of family income and expenditure in Taiwan area*, Executive Yuan, Taiwan, ROC.

Directorate-General of Budget, Accounting and Statistics (DGBAS), 1999c. *National income in Taiwan area*, Executive Yuan, Taiwan, ROC.

Directorate-General of Budget, Accounting and Statistics (DGBAS), 1999d. *Year book of manpower Statistics*, Executive Yuan, Taiwan, ROC.

Directorate-General of Budget, Accounting and Statistics (DGBAS), 1999e. *The trend in multifactor productivity*, Executive Yuan, Taiwan, ROC.

Hanson, K., Rose, A., 1997. Factor productivity and income inequality: a general equilibrium analysis, *Applied Economics* 29, 1061-1071.

Huang, T.Y., Yang, R.J., Chu, Y.H., 1993. On the estimation of energy-related CO2 emissions in Taiwan, *Energy Quarterly* 23(2), 69-82 (in Chinese).

Hsu, G. J. Y., Leung, P., and Ching, C. T.K. 1987, A multi-objective programming and inter-industry model for energy-economic planning in Taiwan, *Energy Systems and Policy* 11, 185-204.

Lee, H., Roland-Holst, D., 1997. The environment and welfare implications of trade and tax policy, *Journal of Development Economics* 52, 65-82.

Parry, Ian W. H., Williams III, R. C., 1999. A second-best evaluation of eight policy instruments to reduce carbon emissions, *Resource and Energy Economics* 21, 347-373.

Perroni, C., Wigle, R.M., 1994. International trade and environmental quality: how important are the linkage?, *Canadian Journal of Economics* 27, 551-567.

Roe, T.L., Yeldan, A.E., 1996. How the doctor's prescriptions may fail: environmental policy under alternative market structures, *METU Studies in Development* 23(4), 577-600.

Yang, Hao-Yen, 2001a. Trade liberalization and pollution: A general equilibrium analysis of carbon dioxide emissions in Taiwan, *Economic Modelling* 19(3), 435-454

Yang, Hao-Yen, 2001b. Carbon emissions control and trade liberalization: Coordinated approaches to Taiwan's trade and tax policy, *Energy Policy* 29(9), 725-734.

APPENDIX. THE EQUATIONS OF THE MODEL

In this appendix is a presentation of the Taiwan's computable general equilibrium model in complete equation form.

A1. Indices

i, j	Production sectors
in, jn	Non-energy sectors and non-fossil-fuels energy sector (i.e. electricity)
ie, je	Fossil-fuels energy sectors (i.e. coal, oil and natural gas)
h	Household income classes

A2. Exogenous Variables and Parameters

$\overline{L}^s$	Aggregate labor supply
$\overline{K}^s$	Aggregate capital supply
$\overline{G}_i$	Government consumption by sector
TR^{GE}	Net total transfer from government to enterprise
TR_h^{GH}	Net total transfer from government to household by income class
TR_h^{WH}	Net total transfer from abroad to household by income class
TR^{WG}	Net total transfer from abroad to government

F_h^L	Net labor income from abroad by income class
F^K	Net capital income from abroad
$INVARB$	Net foreign borrowing
PWE_i	World price of export goods by sector
PW_i	World price of import goods by sector
Ω_i , Γ_i , Φ_i , Z_i	CES production function of efficiency parameter by sector
η_i , ρ_i , τ_i , ϕ_i , ξ_i	CES production function of substitution parameter by sector
ς_i , δ_i , ε_i , ς_i , ζ_i	CES production function of distribution parameter by sector
σ_i^{KL}	CES production function of elasticity of substitution between capital and labor by sector
$\sigma_i^{F \cdot E_{elec}}$	CES production function of elasticity of substitution between fossil fuels and electricity by sector
$\sigma_i^{E_{coal}LF}$	CES production function of elasticity of substitution between coal and liquid fuels by sector
$\sigma_i^{E_{oil}E_{gas}}$	CES production function of elasticity of substitution between oil and natural gas by sector
a_{ij}	Input-output coefficient by sector
$\overline{B}_i$	CES composite goods function of shift parameter by sector
υ_i	CES composite goods function of distribution parameter by sector
ν_i	CES composite goods function of substitution parameter by sector
$\overline{EX}_i$	Export function of shift parameter by sector
e_i	Export function price elasticity by sector
λ_i^w	Labor market distortion parameter by sector
λ_i^r	Capital market distortion parameter by sector
t_h^H	Household income tax rate by income class
β_{ih}^H	Household expenditure shares by sector and income class
s_h^H	Household saving rate by income class
Θ_{ih}^H	Household income distribution shares by sector and income class
s^E	Enterprise saving rate
t^E	Enterprise income tax rate
Θ_h^E	Enterprise income distribution shares by income class
td_i	Indirect tax rate of domestic goods by sector
tm_i	Tariff rate of import goods by sector

θ_i	Capital depreciation rate by sector
π_i	Investment distribution parameter by sector

A3. Endogenous Variables

X_i	Production output by sector
EVA_i	Composition of labor, capital, electricity, and fossil fuels by sector
A_i	Intermediate inputs by sector
VA_i	Value-added by sector
EA_i	Composite of energy inputs by sector
K_i	Capital input by sector
L_i	Labor input by sector
$E_{ie,i}$	Fossil-fuels energy input by sector
$E_{elec,i}$	Electricity input by sector
F_i	Aggregate of the fossil fuels by sector
LF_i	Aggregate of the Liquid fuels by sector
w	Aggregate labor factor price
r	Aggregate capital factor price
U_h	Utility of household consumption by income class
C_i	Household consumption by sector and income class
Y_h^H	Household income by income class
Y^E	Enterprise income
Y^G	Government income
I	Total investment
S^G	Total government saving
Q_i	Composite goods by sector
M_i	Import goods by sector
D_i	Domestic goods by sector
EX_i	Export goods by sector
PVA_i	Net price by sector
P_i	Composite goods price by sector
P_i^F	Aggregate of the fossil fuels price by sector

P_i^{LF}	Aggregate of the liquid fuels price by sector
PD_i	Domestic goods price by sector
PM_i	Import goods price by sector
ER	Exchange rate
PP	GDP deflator
$NGDP$	Nominal gross domestic product
$RGDP$	Real gross domestic product

A4. Equations

A4.1. Production

$$X_{in} = \min\left(A_{in}, EVA_{in}\right) \tag{A1.1}$$

$$X_{ie} = \min\left(A_{ie}, VA_{ie}\right) \tag{A1.2}$$

$$A_{in} = \sum_{jn} a_{jn,in} X_{in} \tag{A2.1}$$

$$A_{ie} = \sum_{j} a_{j,ie} X_{ie} \tag{A2.2}$$

$$EVA_{in} = \Omega_{in}\left(\varsigma_{in} VA_{in}^{-\eta_{in}} + \left(1-\varsigma_{in}\right) EA_{in}^{-\eta_{in}}\right)^{-\frac{1}{\eta_{in}}} \tag{A3}$$

$$VA_i = \Gamma_i\left(\delta_i L_i^{-\rho_i} + \left(1-\delta_i\right) K_i^{-\rho_i}\right)^{-\frac{1}{\rho_i}} \tag{A4}$$

A4.2. Factor Markets

$$L_i = \left(\frac{X_i}{\Gamma_i}\right)^{1-\sigma_i^{KL}} \left(PVA_i \frac{\delta_i}{\lambda_i^w w}\right)^{\sigma_i^{KL}} \tag{A5}$$

$$K_i = \left(\frac{X_i}{\Gamma_i}\right)^{1-\sigma_i^{KL}} \left(PVA_i \frac{1-\delta_i}{\lambda_i^r r}\right)^{\sigma_i^{KL}} \tag{A6}$$

$$PVA_i = \frac{X_i}{\Gamma_i}\left(\delta_i^{\sigma_i^{KL}}\left(\lambda_i^w w\right)^{1-\sigma_i^{KL}} + \left(1-\delta_i\right)^{\sigma_i^{KL}}\left(\lambda_i^r r\right)^{1-\sigma_i^{KL}}\right)^{\frac{1}{1-\sigma_i^{KL}}} \tag{A7}$$

$$\sum_i L_i = \overline{L}^s \tag{A8}$$

$$\sum_i K_i = \overline{K}^s \tag{A9}$$

A4.3. Industrial Energy Demand

$$EA_{in} = \Phi_{in}\left(\varepsilon_{in} F_{in}^{-\tau_{in}} + \left(1-\varepsilon_{in}\right)E_{elec,in}^{-\tau_{in}}\right)^{-\frac{1}{\tau_{in}}} \tag{A10}$$

$$F_{in} = Z_{in}\left(\varsigma_{in} E_{coal,in}^{-\phi_{in}} + \left(1-\varsigma_{in}\right)\left(\zeta_{in} E_{oil,in}^{-\zeta_{in}} + \left(1-\zeta_{in}\right)E_{gas,in}^{-\zeta_{in}}\right)^{\frac{\phi_{in}}{\zeta_{in}}}\right)^{-\frac{1}{\phi_{in}}} \tag{A11}$$

$$\frac{E_{elec,in}}{F_{in}} = \left(\frac{P_{in}^F}{P_{elec}}\right)^{\sigma_{in}^{F\cdot E_{elec}}}\left(\frac{1-\varepsilon_{in}}{\varepsilon_{in}}\right)^{\sigma_{in}^{F\cdot E_{elec}}} \tag{A12}$$

$$\frac{E_{coal,in}}{LF_{in}} = \left(\frac{P_{in}^{LF}}{P_{coal}}\right)^{\sigma_{in}^{E_{coal}LF}}\left(\frac{\varsigma_{in}}{1-\varsigma_{in}}\right)^{\sigma_{in}^{E_{coal}LF}} \tag{A13}$$

$$\frac{E_{oil,in}}{E_{gas,in}} = \left(\frac{P_{gas}}{P_{oil}}\right)^{\sigma_{in}^{E_{oil}E_{gas}}}\left(\frac{\zeta_{in}}{1-\zeta_{in}}\right)^{\sigma_{in}^{E_{oil}E_{gas}}} \tag{A14}$$

$$E_{ie,je} = a_{ie,je} X_{ie,je} \tag{A15}$$

A4.4. Expenditure and Income

$$P_i C_{ih} = \beta_{ih}^H\left(1-s_h^H\right)\left(1-t_h^H\right)Y_h^H \tag{A16}$$

$$Y_h^H = \sum_i \Theta_{ih}^H \lambda_i^w w L_i + F_h^L + \Theta_h^E\left(1-s^E\right)\left(1-t^E\right)Y^E + TR_h^{GH} + TR_h^{WH} \tag{A17}$$

$$P_i\overline{G}_i = Y^G - \sum_h TR_h^{GH} - TR^{GE} - S^G \tag{A18}$$

$$Y^G = \sum_i td_i PD_i X_i + \sum_i tm_i PW_i ER \cdot M_i + t^E Y^E + \sum_h t_h^H Y_h^H + TR^{WG} \tag{A19}$$

$$Y^E = \sum_i \lambda_i^r r \cdot K_i + F^K + TR^{GE} \tag{A20}$$

A4.5. Foreign Exchange Market

$$Q_i = \overline{B}_i\left(\upsilon_i M_i^{-\nu_i} + (1-\upsilon_i) D_i^{-\nu_i}\right)^{-\frac{1}{\nu_i}} \tag{A21}$$

$$M_i = \left(\frac{PD_i}{PM_i}\right)^{\frac{1}{1+\nu_i}} \left(\frac{\upsilon_i}{1-\upsilon_i}\right)^{\frac{1}{1+\nu_i}} D_i \tag{A22}$$

$$EX_i = \overline{EX}_i\left(\frac{ER \cdot PWE_i}{PD_i}\right)^{e_i} \tag{A23}$$

$$\sum_i PD_i EX_i + \sum_h F_h^L + F^K + \sum_h TR_h^{WH} + TR^{WG} = \sum_i PW_i ER \cdot M_i + INVARB \tag{A24}$$

A4.6. Prices

$$PM_i = PW_i ER \cdot (1 + tm_i) \tag{A25}$$

$$P_i = \frac{1^{\bullet}}{\overline{B}_i}\left(\nu^{\frac{1}{1+\nu}} PM_i^{\frac{\nu}{1+\nu}} + (1-\nu)^{\frac{1}{1+\nu}} PD_i^{\frac{\nu}{1+\nu}}\right)^{\frac{1+\nu}{\nu}} \tag{A26}$$

$$P_{in}^F = \frac{1}{Z_{in}}\left(\varsigma_i^{\sigma_i^{E_{elec}LF}} P_{coal}^{1-\sigma_i^{E_{elec}LF}} + (1-\varsigma_i)^{\sigma_i^{E_{elec}LF}} P_{in}^{LF\,1-\sigma_i^{E_{elec}LF}}\right)^{\frac{1}{1-\sigma_i^{E_{elec}LF}}} \tag{A27}$$

$$P_{in}^{LF} = \left(\zeta_i^{\sigma_i^{E_{oil}E_{gas}}} P_{oil}^{1-\sigma_i^{E_{oil}E_{gas}}} + (1-\zeta_i)^{\sigma_i^{E_{oil}E_{gas}}} P_{gas}^{1-\sigma_i^{E_{oil}E_{gas}}}\right)^{\frac{1}{1-\sigma_i^{E_{oil}E_{gas}}}} \tag{A28}$$

$$PP = \frac{NGDP}{RGDP} \tag{A29}$$

$$NGDP = \sum_i \Theta_{ih}^H \lambda_i^w wL_i + \sum_i \lambda_i^r rK_i + \sum_i td_i PD_i X_i + \sum_i PP \cdot \theta_i K_i + \sum_i PW_i \cdot ER \cdot tm_i M_i \tag{A30}$$

$$RGDP = \sum_h C_{ih} + I + \sum_i \overline{G}_i + \sum_i \frac{PD_i EX_i}{P_i} - \sum_i \frac{PW_i ER \cdot M_i}{P_i} \tag{A31}$$

A4.7. Market Equilibrium

$$Q_i = \sum_j a_{i,j} X_j + \sum_h C_{i,h} + \pi_i I + \overline{G}_i; \quad i \in \text{non-energy sectors} \tag{A32.1}$$

$$Q_i = \sum_j E_{i,j} + \sum_h C_{i,h} + \pi_i I + \overline{G}_i; \quad i \in \text{energy sectors} \tag{A32.2}$$

$$X_i = EX_i + D_i \tag{A33}$$

A4.8. Macro Closure Rule

$$I = \sum_i \theta_i K_i + s^E\left(1 - t^E\right)Y^E + \sum_h s_h^H\left(1 - t_h^H\right)Y_h^H + S^G - INVARB \tag{A34}$$

Chapter 3

STRATEGIES FOR CARBON TAXATION FOR REDUCING CO_2 EMISSIONS IN JAPAN AND KOREA

Toshihiko Nakata
Tohoku University
Graduate School of Engineering
Aoba-Yama 01
Sendai 980-8579, JAPAN
Tel: +81 (22) 217-7004
FAX: +81 (22) 217-7004
E-mail: nakata@cc.mech.tohoku.ac.jp

ABSTRACT

This paper focuses on the impacts of using carbon and energy taxes to reduce carbon emissions from the energy systems of both Japan and Korea. Two countries have few domestic fossil fuels, resulting in shaky condition in current energy supply. Ensuring energy security is a national issue once again to stabilize economic development and to meet Kyoto protocol at a reasonable cost. Besides, the Japanese are increasingly keen to reduce the use of such 'dirty' fossil fuels as coal and oil in favor of gas.

A partial equivalent model of the national energy sector has been developed to forecast changes in the energy system out to the year 2040. The model accounts for the changes in energy technology capacities, fuels, and consumption in response to policy initiatives, such as taxes. We find that carbon and energy taxes will decrease carbon dioxide emission to a proposed target. The total cost in terms of supplying energy will be similar for either approach. However, the model also indicates that carbon taxes cause a shift in resources used from coal to gas. Since energy security is a primary concern to Japan and Korea, maintaining a diverse base of resources is very important. Policies that would eliminate coal, and efficient coal based technologies, may not be desirable. The development of clean coal technologies and advanced transportation technologies suitable for national energy systems should be the next target to overcome the limit of carbon taxes. Another policy option for the reduction of CO_2 emissions is the Japan-Korea cooperation in energy technology. Coupling of utilities such as gas pipeline and electricity grid will have a great impact on ensuring energy security for Far

East countries. Though the deregulation of energy markets is still on the way, cost effective and environmental sound strategies should be integrated to show up-to-date energy scenarios.

1. Introduction

The Kyoto conference calls for substantial reductions in the emissions of carbon from the world's energy systems. However, it is not clear how these reductions should be achieved. Each country has its own unique considerations of energy resources, economic demand patterns, and energy security concerns. Thus, the best strategy for each country will be different. In this paper we examine possible policy approaches for Japan and Korea. There are two questions that must be addressed for these countries: First, how can the energy system be configured so as to meet future CO_2 constraints, Second: what sorts of policies could be used to encourage the system into new configurations? The possible policies must be evaluated in light of their potential cost, emissions other that greenhouse gasses, and implications for national energy security.

Since almost all the energy resources are imported in Japan and Korea, energy security should have some priority over environmental constraints. Integrated analysis, which includes energy and environmental models, will make an important role to find out the most suitable scenario for national energy systems. Japan and Korea must assess different approaches to controlling CO_2 emissions. The impacts of controls on CO_2 emission from electricity in Canada and the USA have been studied (for example, Chung et al., 1997), however there are few papers concerning electric power reconfiguration in Japan and Korea.

The transportation sector has the characteristics of mostly using petroleum. Electric trains are powered indirectly from a variety of energy sources; however, their share of energy consumption in transportation was no more than 2.3 percent in Japan in the year 1997. Even when the cost of petroleum rises, conventional vehicles cannot switch right away to alternative energy sources. Hybrid vehicles, though still using gasoline as a fuel, could be one alternative technology in the transportation sector. Such vehicles have excellent fuel efficiency and a strong likelihood of drastically reducing CO_2 emissions.

Reaching the goals of emission reductions will require some action on the part of the government. There are a number of policy actions available to the government, ranging from regulatory incentives to taxes. Moreover, there are many energy alternatives in combination with technologies for energy conversion and with market forces. For example, the electric sector has more choices of energy resources than the transportation sector, in which only petroleum can meet most of the demand today. Electricity can be powered by fossil fuels, nuclear energy, and renewables such as hydro, wind and solar energy.

Nuclear energy has been an important energy source since the first oil embargo occurred in the early 1970s. Energy security has been a first concern for policy makers in Japan. Japan has increased its reliance on nuclear power because it has limited domestic energy resources. At present, 51 commercial nuclear power plants are in operation, and the government had wanted to build 10-13 more by the year 2010 (JPET, 1998; Kiso, 1998 and Matthews, 1998). Recently, the Ministry of Economy, Trade and Industry in Japan (METI) announced that nuclear power stations could be technically in use for sixty years under strict maintenance condition. Nuclear power has an advantage over fossil power stations in terms of CO_2

emissions. The government maintains that nuclear power helps to prevent global warming because nuclear power plants do not emit CO_2. In fact, as the nuclear power reactors operate past the period of paying off their capital costs, they produce electricity cheaper than any other energy source, because their operations and maintenance costs are the lowest. The capacity factors and safety records of power reactors have improves dramatically worldwide in recent years. In addition, the construction time and costs of power reactors have been improved, e.g., the Tokyo Electric/GE advanced 1,300 MWe BWR was built in just four years and within budget. By the year 2040 it becomes possible that the fast breeder reactors and high temperature helium gas cooled reactors will be built for a number of reasons. This will increase the efficiencies of electricity generation and lower the CO_2 emissions significantly, particularly in process of heat applications as well as in electricity generation.

However, current changes in ESI toward deregulation and privatization make electric power utilities conservative about future investment in nuclear power stations in Japan (OECD, 2001). The Japan's long-term recession forces investors to assume the risk of their investments. Independent Power Producers (IPP), which are recently allowed to come into the electricity market under competition, prefer combined-cycle power stations which use liquefied natural gas as a fuel. Combined-cycle plants have priority over nuclear power: Initial and capital costs are much lower, and gas turbines allows flexibility to meet the demand changes. Total heat efficiency of advanced combined-cycle exceeds 50%, which saves fuel cost clearly.

Depletion and environmental effects of burning fossil fuels such as oil, gas and coal is also serious (Campbell and Laherrere, 1998; Banks, 2000 and Gan, 1998). External costs of atmospheric pollution are not included in the price of fossil fuels correctly. The irrationality of burning oil and gas rather than preserving them for the production of petrochemicals, etc, is not well discussed. The governments are compelled to find a solution to the problem.

The issue of nuclear safety has top priority at all time and economic aspects have an important role in developing nuclear energy policy. The question of what would be the best way to control the emission of CO_2 remains unsettled (US Department of Energy, Energy Information Administration, 1998a; Sato et al., 1998; Zongxin and Siddiqi, 1995; Romerio, 1998 and Bodde, 1998). The problem, which we have to consider next, is an integrated concept of national security to meet the environmental constraints. Nuclear power needs to be examined in detail now from the standpoint of what we call environmental security (Allenby, 2000; Kagramanian et al., 2000 and Kaya, 1997). Whereas we recognize the importance of nuclear power, it seems that its role from economic and environmental aspects in the future is not well organized in Japan. The point of Kyoto and greenhouse emissions is not to give nuclear power a free ride to bright future (Doucet, 1999).

This paper uses a multi-period market equilibrium model to explore the cost and energy security implications of carbon taxes as policies for reconfiguring the national energy system, for all sectors including electric power. Moreover, we explore the impact of hybrid vehicles on CO_2 emissions in the Japanese energy system. A BTU tax is hypothetically imposed to evaluate the effect of fuel cost on energy consumption in the transportation sector and on CO_2 emissions. Financial parameters such as capital costs and operating costs are included to estimate the profits for taxation purposes. The effect of factors such as capital investments, resource depletion, and the market penetration of advanced technologies were included, in addition to a multi-period modeling of the price-quantity equilibrium in various markets. The impact of regulatory policies involving devices such as taxes on emissions or constraints on quantities and prices has been modeled.

The role of nuclear power to reduce CO_2 emissions is also examined. To make it clear, nuclear power phase-out (Nordhaus, 1997; Andersson and Haden, 1997; Welsch, 1998; Hoster, 1998; Welsch and Ochsen, 2001 and Haas and Auer, 2001) is hypothesized in the model to evaluate the difference in CO_2 emissions. The carbon tax is tentatively introduced to stress the increases in fossil fuel consumption. An economic aspect of two cases, with or without nuclear power, is also examined.

Another policy option for the reduction of CO_2 emissions is the Japan-Korea cooperation in energy technology. We focus on the energy systems in Korea to see the possible options for reconfiguring energy networks between two countries in East Asia.

2. Impacts of Carbon and BTU Taxes on Energy Systems in Japan

2.1 Policy Approaches to Reduce CO_2 Emissions

Reducing carbon emissions will require some action on the part of the government. In recent years many studies on the cost of CO_2 emission abatement have been published (Richels and Sturm, 1996; Schultz and Kasting, 1997). There are a number of policy actions available to the government ranging from regulatory incentives to taxes. Here we outline a few of these possibilities.

The Government can encourage the construction of certain types of technologies through the permitting processes and financial encouragement. It has requested that Japan's electric power utilities construct twenty nuclear power stations between now and 2010 which means approximately a fifty percent increase in nuclear power generation capacity. It would be unrealistic for electric power utilities to build more nuclear power stations at this time without government encouragement because of its higher capital cost and the difficulty of obtaining public acceptance at local sites.

Another alternative is implementation of a carbon tax. A carbon tax will raise the price of high carbon fuels, such as coal, and lead to a greater reliance on low carbon fuels. Since this tax directly discourages carbon use, it is expected that will lead to the most efficient reconfiguration of the energy system to meet emissions limits (Environment Agency of Japan, 1998). However, coal appears to be one of the most abundant fossil fuels. From the viewpoint of energy security, it is expected that coal, especially when used with clean coal technologies, should have an important role in Japan's energy mix. In this sense a carbon tax may not be the best way to meet Japan's objectives. An energy tax applied to all fossil fuels could be another approach for meeting both the CO_2 emission constraint and providing a good mix of energy resources.

Because there are many alternative technologies and fuels available to meet emission targets, particularly in the electric sector, it is important for policy makers and engineers to understand the implications of different policies intended to meet the targets. The model that we discuss here includes effects such as the market penetration of advanced technologies, capital investments, capacity retirements, and resource depletion. The impact of regulatory policies can be modeled through devices such as taxes on emissions or constraints on quantities and prices.

2.2 National Energy Model for Japan

Modeling has been a tool for national energy planning since the mid-1970s. At that time models were used to understand the implications and means of coping with the first oil embargo. More recently global warming has become one of the most significant topics for decision-makers. Just as physical models can predict the impact of increased CO_2 on climate, energy-economic models can show the economic and technical impacts of alternative economic strategies to minimize emissions.

There are previous analyses of the Japanese energy system. An outlook for Japan's energy supply and demand is published (for example, Kibune and Kudo, 1996). By using the CRIEPI model, cost effectiveness of CO_2 emission controls were studied (Yamaji et. al., 1993). In this study a macro input-output model was used to predict the impact of subsidies on CO_2 emissions in Japan. However, the model included little in the way of market functions in each sector and energy prices were derived explicitly.

Impacts of carbon taxation on Japan's economy as a whole were examined by using general equilibrium model (Goto, 1995). Macro economic costs of CO_2 emission control were also studied (Goto and Sawa, 1993). In their model the industrial sector was divided into twelve sub-sectors. However, the electricity sector was treated as roughly one single sub-sector, making it difficult to estimate the effects of such policies on the structure of the electric sector. There are few analyses of Japan's energy system that disaggregate the electric generating sector so that the impacts of such policies on technologies and fuel use can be examined.

An Energy-Economic Model

An energy economic model for evaluating CO_2 emissions policies should incorporate energy technologies, economics, and environmental impacts. Technology innovation and its efficiency improvement are factors that should be included for this model as well.

There are several modeling approaches available in energy economics based on either optimization or equilibrium approaches shown in Table 1. We have used the META•Net economic modeling system developed at Lawrence Livermore National Laboratory (Lamont, 1990). META•Net is a partial equilibrium modeling system that allows for explicit price competition between technologies, and can constrain or tax emissions. It allows a user to build and solve complex economic models. It will construct an energy model as a network of processes such as end-uses (price sensitive demands), markets (which allocate market shares based on relative prices), conversion processes (which compute inputs required to meet output requirements based on efficiencies, and compute prices based on capital and operating costs), and resources (which can be exhaustible, or can follow a set price track).

Table 1: Comparison of META•Net to existing modeling approaches

Modeling technologies	Tools	Developer	Feature
Fixed coefficient model	LEAP[a]	Tellus Institute	All impacts and inputs are set in a fixed ration to the out put required
Technology based market model	MARKAL[b]	BNL	Rigid structure of linear programming
Technology based market model	META•Net[c]	LLNL	Solves the economic equilibrium problem directly
General equilibrium model	Second generation model[d]	PNL	World scale modeling
General equilibrium model	GREEN[e]	OECD	World scale modeling

[a]Lazarus, M. Heaps, C. and Raskin, P. (1997). LEAP Long-range energy alternatives planning system. Stockholm Environmental Institute, Boston, Massachusetts.

[b]Lydick, J. Morris, S. C. Lee, J. and Goldstein, G. (1990). Demo MARKAL Abbreviated Version of the US MARKAL Energy Systems Model, Created for Demonstration: Abilities, Limitations, and Demonstration, BNL-47782, Brookhaven National Laboratory, Upton, New York

[c]Lamont, A. (1994). User's Guide to the META•Net Economic Modeling System Version 1.2. UCRL-ID-122511, Lawrence Livermore National Laboratory, Livermore, California.

[d]Edmonds, J. A. Pitcher, H. M. Barns, D. Baron, R. and Wise, M. A. (1995). Modeling future greenhouse gas emissions: The second generation model description. In Modeling global change, ed. Klein L. R. and Lo, F., pp. 295-362. United Nations University Press

[e]Lee, H. Oliveira-Martins, J. and Mensbrugghe, D. (1994). The OECD Green Model: An Updated Overviews. Technical paper No. 97, OECD, Paris.

Commodities flow through this network from resources, through conversion processes and market, to the end-users. META•Net finds the multi-period equilibrium prices and quantities. The solution includes the prices and quantities demanded for each commodity along with the capacity additions (and retirements) for each conversion process, and the trajectories of resource extraction.

Although the changes in the economy are largely driven by consumers' behavior and the costs of technologies and resources, they are also affected by various governmental policies. These can include constraints on prices and quantities, and various taxes and constraints on environmental emissions. META•Net can incorporate many of these mechanisms and evaluate their potential impact on the development of the economic system.

The Japan model which we have designed is shown in Figure 1. It has sixty-four processes including eight demand nodes in the industrial, commercial, residential and transportation sectors, eight resource nodes modeling purchases of coal, natural gas, petroleum and nuclear fuel on the world markets, along with domestic hydropower and other renewables. Additional processes model the conversion of fuels to heat, electrical services, and transportation services. Necessary operating parameters such as capital cost and operating cost are used based on the current references (The Federation of Electric Power Companies, 1998: US Department of Energy, Energy Information Administration, 1998b: MITI, Agency of Natural Resources and Energy, 1998). Table 2 shows capital and operating cost used for this study in the electricity sector.

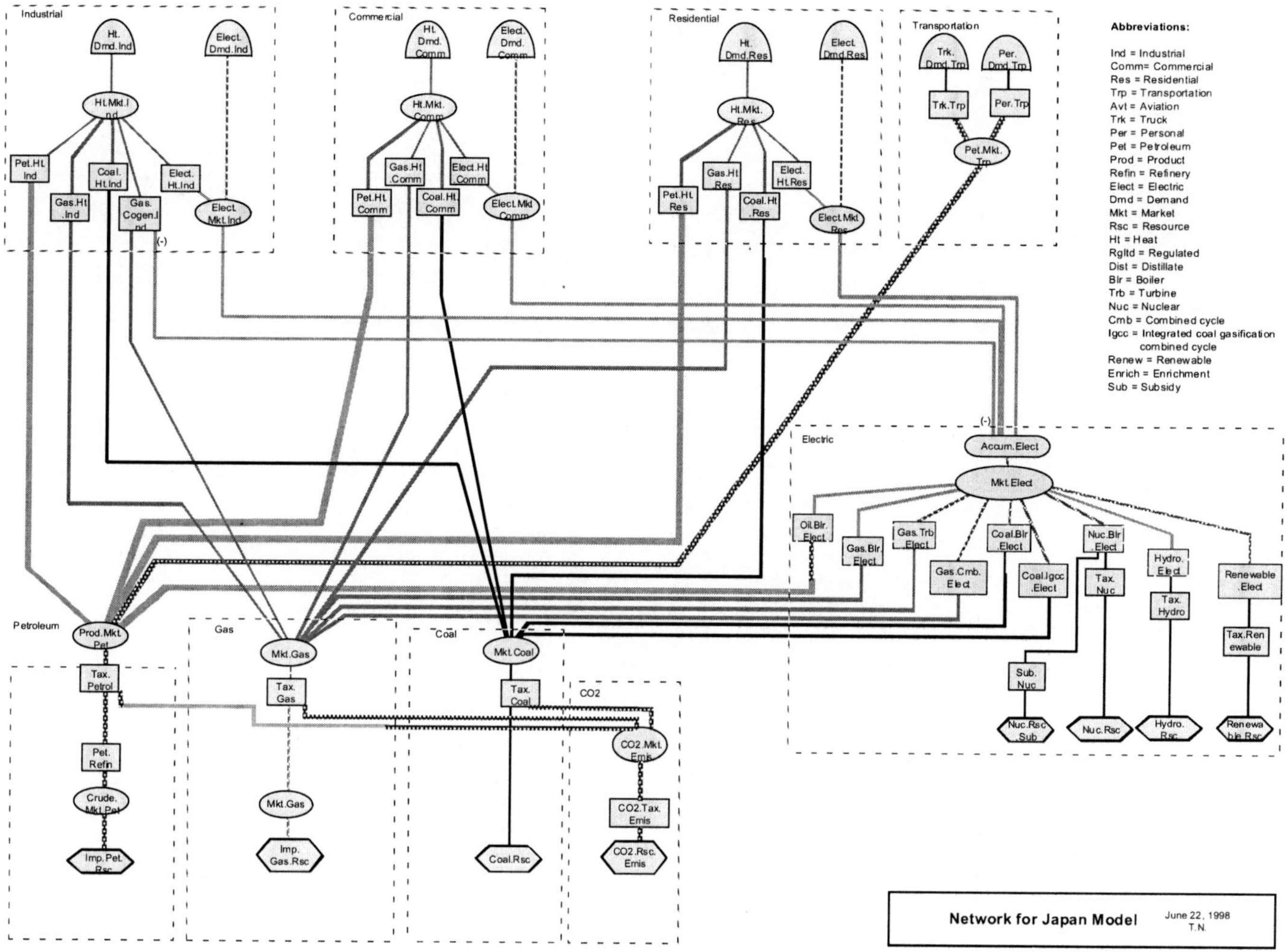

Figure 1: Network for Japan model

Table 2: Capital and operating cost in electricity sector

Electric power generation	Capital cost $/(mmBTU/Yr)	Operating cost $/mmBTU
Oil Boiler	49	9
Gas Boiler	51	10
Gas Turbine	30	10
Gas Combined Cycle	47	11
Pulverized Coal	68	14
Coal IGCC	110	13
Nuclear Boiler	80	16
Hydro	154	1
Renewable	50	4

Sources:

- The Federation of Electric Power Companies. (1998). *Handbook of Electric Power Industry,* Tokyo.
- US Department of Energy, Energy Information Administration. (1998). *Assumptions to the Annual Energy Outlook*, Washington DC.

Key Parameters and Assumptions

There are several key assumptions that drive any analysis of this type. These include growth rates, government policies to encourage particular technologies, and demand response to changes in price. We have assumed a moderate rate of growth over the model horizon. Table 3 presents assumptions about the growth rates and demand elasticities in each sector.

Due to government policies to maintain a nuclear generation capacity, it is assumed that the share of Nuclear Boiler power generation will remain constant at its 1995 level (about 35% of power generation output) until 2040. Hydropower is also set at constant generation quantity because of physical constraint on siting additional capacity. The Oil Boiler has maximum quantity constraint in accordance with the IEA agreement made just after the first oil embargo. New technologies such as coal integrated combined cycle and renewable power will be available in 2005. Their capacities are allowed to increase ten percent every year commencing 2005. Aviation transportation demand has not been included in this model since it is quite small in Japan.

Table 3: Growth rate and elasticity assumptions for end-use sector

Process	Abbreviated name	Annual demand in starting period[a]	Annual rate of demand growth (at a constant price)[a]	Demand elasticity[b]
Industrial heat demand (mmBTU/yr)	Ht.Dmd.Ind	5.10E+09	0.002	-0.34
Industrial electricity demand (mmBTU/yr)	Elect.Dmd.Ind	1.24E+09	0.008	-0.34
Commercial heat demand (mmBTU/yr)	Ht.Dmd.Comm	9.00E+08	0.013	-0.24
Commercial electricity demand (mmBTU/yr)	Elect.Dmd.Comm	6.64E+08	0.020	-0.24
Residential heat demand (mmBTU/yr)	Ht.Dmd.Res	1.10E+09	0.006	-0.30
Residential electricity demand (mmBTU/yr)	Elect.Dmd.Res	7.37E+08	0.020	-0.30
Truck transportation demand (Ton-km/yr)	Trk.Dmd.Trp	2.14E+08	0.003	-0.17
Personal transportation demand (passenger-km/yr)	Per.Dmd.Trp	6.43E+08	0.010	-0.23

[a]The Energy Data and Modeling Center. (1998). *EDMC Handbook of Energy & Economic Statistics in Japan*, Tokyo.
[b]Based on the data by Nagata, Y. (2000, June 21). Personal communication.

2.3 Policy Scenarios Analyzed

The study focuses on a target rate of emissions of around 400 million tons of carbon (mmTC) by the year 2040. This is higher than the target of 280 mmTC in year 2010 that was specified by the Kyoto protocol (MITI, Agency of Natural Resources and Energy, 1998). Within this analysis, reaching the level of 280 mmTC appears to be difficult and might not be realistic for Japan. The discussion below outlines some of the issues raised by large carbon reductions. For most of the analyses, we have used a less stringent target.

The study evaluates the effects of carbon and energy taxes. In these cases we assume that government regulatory encouragement will keep the level of nuclear capacity at around 35 percent of the electric demand. Our investigation first developed a Reference Case analysis that showed the investments in energy technologies, resource use and emissions in the absence of policy actions. We then determined the level of tax needed to reach the required level of carbon emissions in the year 2040. We evaluated the cost of that approach and the configuration of the energy systems that results from it.

Reference Case Scenario

Under this scenario it is assumed that there is no intervention in the energy system.

Carbon Tax Scenario

A carbon tax is expected to be the most efficient approach to reducing carbon emissions. It has been already implemented in Sweden, Finland, Norway, Denmark and Holland in range of \$200/tonC (Sawa, 1997). In our study we evaluated several different tax rates to determine the rate that achieved the target reduction in the year 2040.

The taxes were introduced gradually over time, increasing the tax rate in uniform steps each period until the maximum rate in 2040 was reached. This avoids the dislocation caused by a sudden, drastic shift in the tax rate.

Energy Tax (BTU Tax) Scenario

An energy tax is another way to reduce emissions. In our study only fossil fuels such as petroleum, gas and coal are charged. Again, the tax rate was gradually increased over time.

2.4 Discussion for Taxation Scenarios

The discussion in this section highlights the changes in the energy system that result from reducing carbon emissions through carbon and energy taxes, and the costs of implementing these changes. Some figures in this section summarize the results from the year 2040, the last year of the analysis.

As the results show in Figure 2, the carbon tax suppresses the increase of CO_2 emissions. Without any taxation CO_2 emission will increase until reaching 491 mmTC in 2040. When the carbon tax is introduced, the increase of CO_2 emissions is strongly suppressed. At \$160/tonC, CO_2 emission will reach 391 million tons of carbon in 2040, which corresponds to a reduction of 100 mmTC.

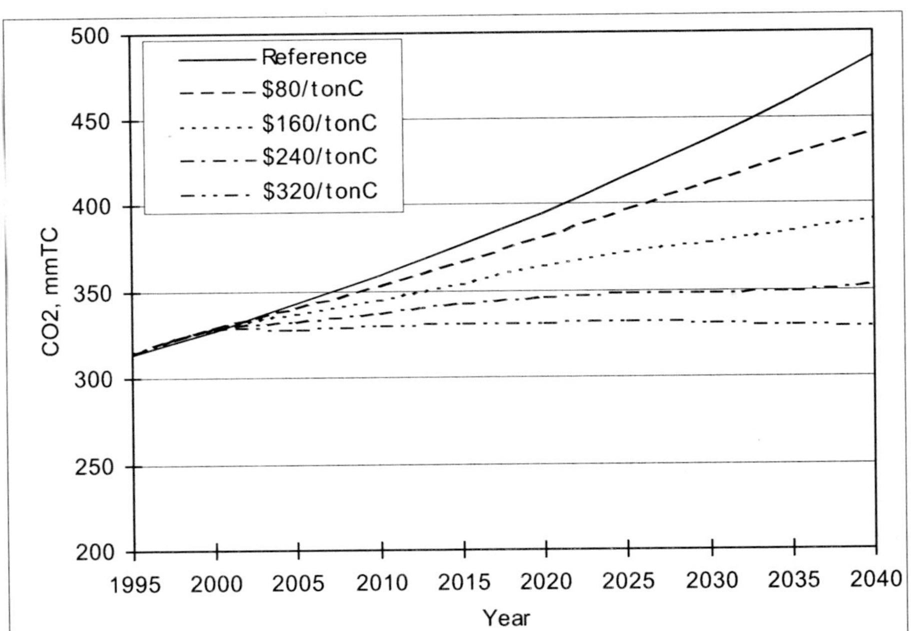

Figure 2: Impact of carbon taxes on CO_2 emissions

Using a similar analysis, we found that an energy tax of $4.5/mmBTU reduced carbon emissions to 400 million tons of carbon in 2040.

Changes in the System Leading to Reduction in Carbon Emissions

The taxes cause a reduction in carbon emissions through adjustments to the system. The costs of these adjustments are important for policy making. However, the *nature* of the structural changes may be equally important for Japan. In this section we discuss the structural changes and the costs resulting from the two taxes.

Figure 3 illustrates the development of the electric power sector to the year 2040 under the reference case. The quantities of coal boiler and gas power show steady increases corresponding to annual demand growth. Clean coal technologies such as Integrated Coal Gasification Combined Cycles (IGCC) are introduced to some extent. The quantities of nuclear power and hydroelectric power are set at a constant level based on the assumptions which we have made for this analysis.

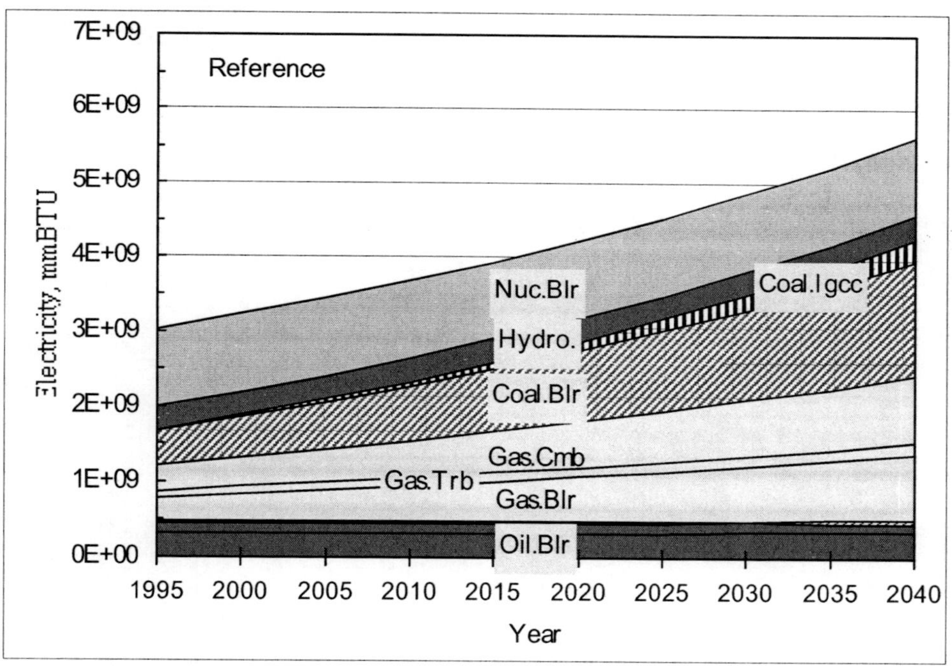

Figure 3: Changes in electric power generation for the reference case

Figure 4 shows the changes in electric power generation under carbon tax of $160/tonC. Under this tax, the coal boiler output decreases by about two-thirds (in 2040) while the output of the gas boiler and gas-combined cycle increases by about a quarter. IGCC does not penetrate the market in this tax case. Renewable energy does penetrate the market after about 2020.

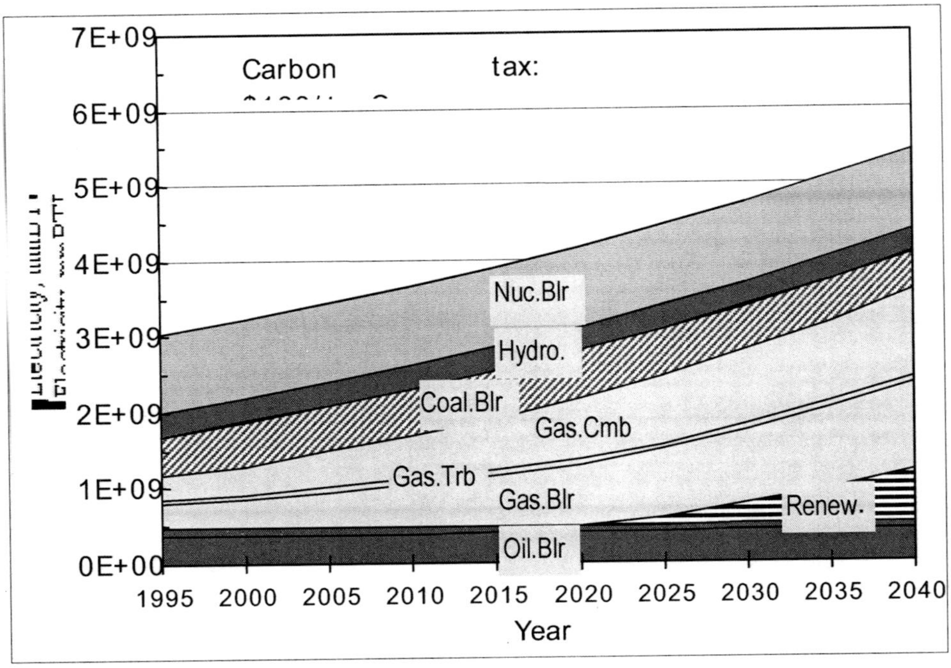

Figure 4: Changes in electric power generation for the carbon tax case

Figure 5 shows the changes in electric power generation under energy tax of $4.5/mmBTU. Under this tax, the coal boiler output as well as the gas boiler and gas-combined cycle decreases by about one-third (in 2040).

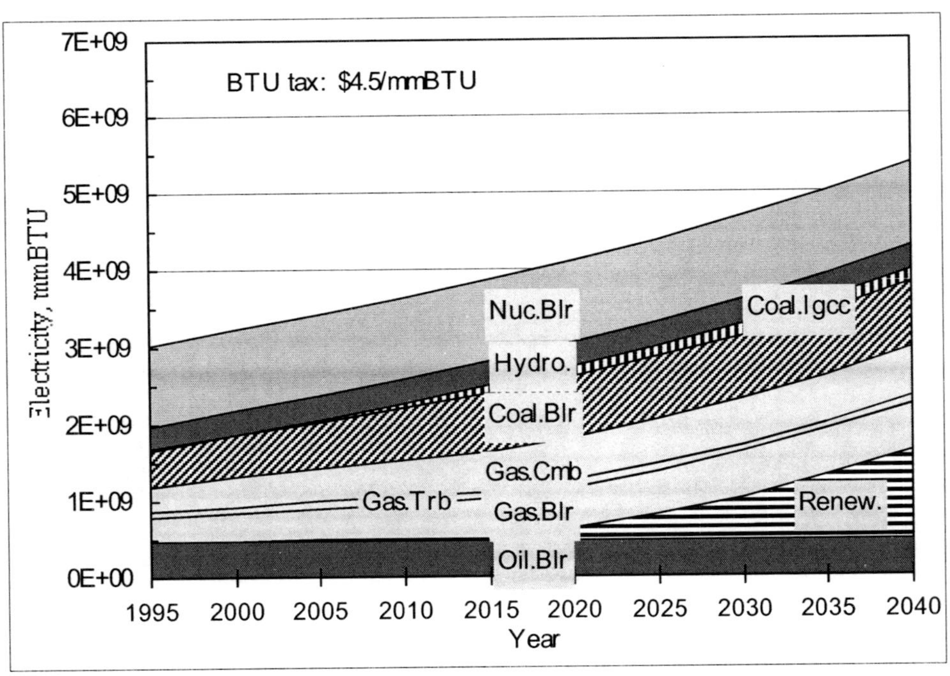

Figure 5: Changes in electric power generation for the energy tax case

Carbon emissions are also reduced through changes in the total energy services required. Figure 6 illustrates the changes in end-use energy under the two tax scenarios compared to the base case scenario. It shows that the industrial heat and residential heat end-use sectors have the largest percentage reductions (6 to 12 percent reduction). Industrial heat is the largest single category of end use, so the change in that sector accounts for a substantial amount of the carbon reduction. Such a reduction in energy use in these sectors could be achieved through improved efficiency, or shifting to less energy intensive products. On the other hand, personal transportation and truck transportation end-use sector have the smallest percentage reductions (1 to 2 percent).

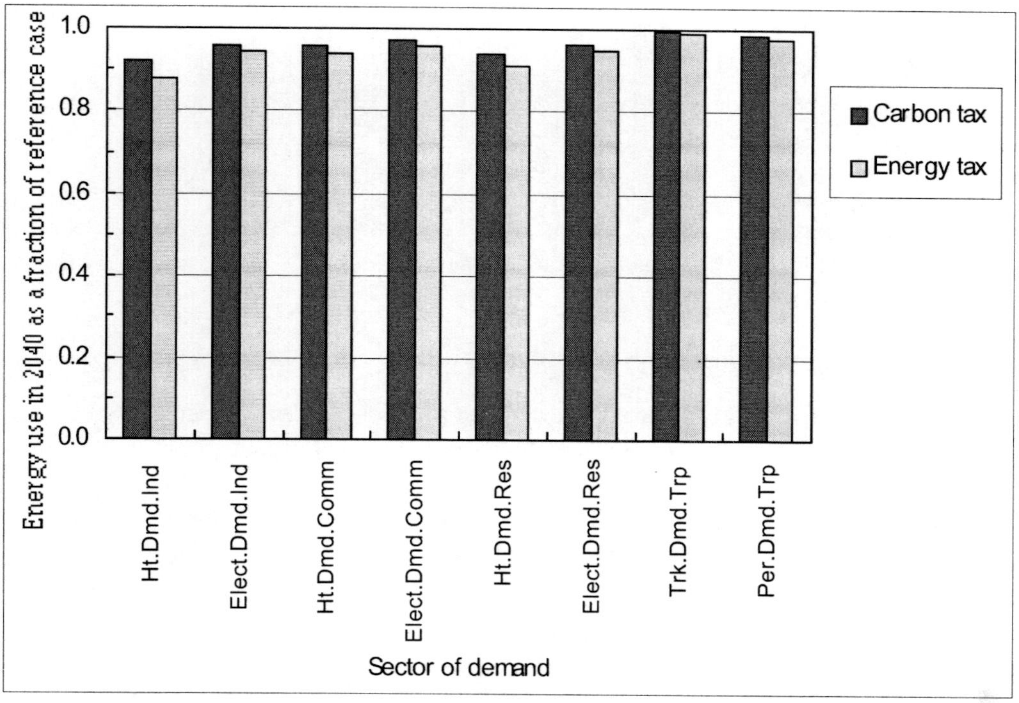

Figure 6: Energy used in each end-use sector under two scenarios in year 2040

Primary energy consumption for the three scenarios is shown in Figure 7. Under both scenarios coal consumption decreases, and natural gas increases under the carbon tax and decreases under the energy tax. The carbon tax greatly reduces coal consumption — coal is nearly replaced by natural gas under the carbon tax. Thus under the carbon tax, one major energy resource is nearly eliminated from Japan's energy mix. Petroleum decreases somewhat, but still maintains the largest share which is similar in both the tax cases.

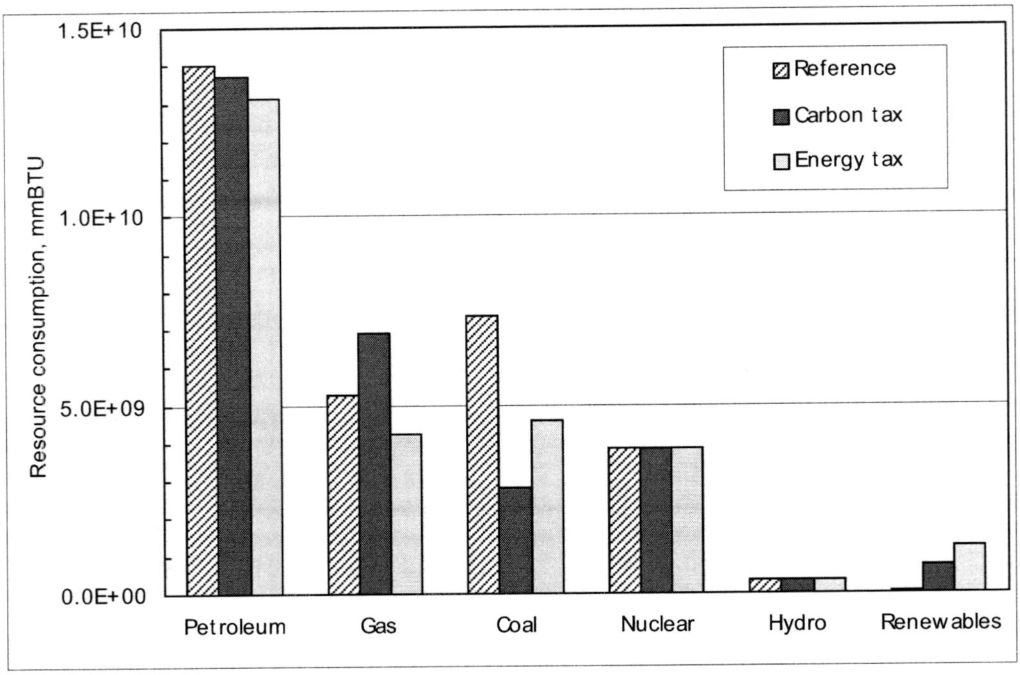

Figure 7: Primary energy used for two the scenarios in year 2040

Under the tax scenarios neither tax was applied to renewable energy. Consequently, both taxes encourage use of renewables in electric power generation.

Table 4 shows the CO_2 emissions from fossil fuels for both reference and taxation scenarios in year 2040. The emission characteristics are similar to fossil fuels consumption shown in Figure 7. As the tax rate becomes higher, CO_2 from coal combustion decreases drastically and CO_2 from gas increases. The amount of CO_2 changes in gas is less than the amount of CO_2 changes in coal. Changes for petroleum is relatively small compared with those of two other fossil fuels.

Table 4: CO_2 emissions from fossil fuels for reference and taxation scenarios in year 2040. Unit: mmTC

	Reference	Carbon tax $160/tonC	BTU tax $4.5/mmBTU
Petroleum	230	223	213
Gas	74	98	61
Coal	187	70	114
Total	491	391	388

Financial Impacts of the Taxes

A summary of the consumer costs and tax revenues is shown in Table 5. Total energy consumption for each case is also shown there. There is a reduction of total energy use in both tax cases. This is caused by the increase in prices due to the taxes, which results in less demand in each end-use sector.

Table 5: Summary of costs and tax revenues for the five scenarios. The costs are the discounted present values of the total stream of costs over the time horizon. The discount rate was 5%. Unit: US$ billion

	Reference	Carbon tax $160/tonC	BTU tax $4.5/mmBTU
Capital cost	2.12E+03	2.10E+03	2.14E+03
Operating costs	5.27E+02	5.12E+02	5.14E+02
Fuel costs	3.70E+02	4.31E+02	3.57E+02
Total direct energy costs	3.02E+03	3.04E+03	3.01E+03
Tax revenue	0	6.30E+01	0
Total cost to consumers	3.02E+03	3.11E+03	3.10E+03
Total energy consumption in 2040 (Quads)	30.8	28.4	27.4

The cost calculations in Table 5 show that both taxes cost the consumer about the same amount for the cost of capital, operations, and fuel in the energy sector. We do not include the tax itself as a cost since it would offset other taxes. In these figures the total cost with the energy tax is slightly less than the cost with the carbon tax. This might appear to be anomalous since the carbon tax is assumed to be more efficient. However, under the energy tax, part of the reduction in carbon is achieved through the reduction of energy services, presumably through greater efficiencies in end uses or by foregoing services. The costs of those measures are not accounted for here. In addition, we note that the total tax revenues under the carbon tax are about one third less than under the energy tax scenario. Thus a dollar's worth of carbon tax is far more effective in reducing carbon emissions than a dollar's worth of energy tax.

3. Impacts of Hybrid Vehicles on Energy Systems in Japan

3.1 Policy Scenarios Analyzed

In this section we see the transportation sector which still shows demand increases for decades in Japan. In order to understand technology innovation and its impacts on CO_2 emissions, hybrid vehicles are taken into account for our energy-economic model. It is assumed that hybrid vehicles will be available for passenger transportation in 2005 in a particular maximum quantity, which will increase ten percent every year commencing in 2005. Based on the reference (Energy Data and Modeling Center, 1998) the performance factors of both conventional and hybrid vehicles are shown in Table 6.

In this investigation a reference case analysis was first developed to model the expected emissions, resource use, and investments in energy technologies in the absence of policy actions. The next step was to model the effects of imposing a BTU tax with the goal of a reduction of 100 mmTC by the year 2040.

Table 6: Performance of vehicles

Type of passenger vehicles	Hybrid[a]	Conventional[b]	*Hybrid/Conventional*
Specific capital cost $/(mmBTU/Yr)	824	687	*1.2*
Ancillary operating cost $/mmBTU	27.46	34.33	*0.8*
I-O coefficient (mmBTU/K passenger mile)	1.431	3.578	*0.4*

[a]Based on Hybrid Car PRIUS, (1999) TH0015-9903, Toyota Motor Corporation, Nagoya.
[b]Based on Energy Data and Modeling Center. (1998). *EDMC Handbook of Energy & Economic Statistics in Japan*, Tokyo.

In the reference case scenario it is assumed that there is no intervention in the energy system, except that nuclear power is subsidized as needed to maintain a 35 percent share of the electric market (the share in 1995). Under this scenario, CO_2 emissions rise to 491 mmTC by the year 2040.

Next, the impact of hybrid vehicles on the transportation sector is examined. A BTU tax is expected to be one of the most efficient approaches to reduce carbon emissions (Sawa, 1997). This study evaluated several different tax rates to determine the rate that would achieve the target reduction in the year 2040. Both cases – with and without BTU tax – were extrapolated to explore the difference in fuel prices. The cost of each approach and the configuration of energy systems that results from it are evaluated.

In the taxation case, the BTU tax was assumed to be imposed in increments, starting at zero in the first period and then rising by a fixed amount in each subsequent period until reaching the target amount of the tax. This approach avoids the sudden shock to the system and the stranded assets that can result from rapid market changes.

It was found that a tax rate of $4.5/mmBTU produced an emission rate of 388 mmTC in 2040, achieving a reduction of 100 mmTC in the emission rate. The tax was introduced gradually over time, increasing the tax rate in uniform steps in each period until the maximum rate was reached in 2040.

In the scenarios involving hybrid vehicles, both cases – with or without BTU taxes – were again studied. In the reference case without BTU taxes, CO_2 emissions rise to 472 mmTC by the year 2040. With a BTU tax at the rate of $4.5/mmBTU, CO_2 emissions rise to 350 mmTC by the year 2040. Again, the tax rate was gradually increased over time.

3.2 Discussion for the Passenger Transportation Sector

The discussion in this section highlights the changes in the energy system that result from reducing CO_2 emissions through the use of hybrid vehicles and a BTU tax, and the costs of implementing these changes.

First, the changes in the passenger transportation sector from introducing hybrid vehicles in 2005 are illustrated in Figure 8. We can see a way in which the system has adjusted to the introduction of hybrid vehicles in 2005. There is an increase in the use of hybrid vehicles from the year 2005 until 2040. The share of hybrid vehicles eventually reaches 50 percent in 2040.

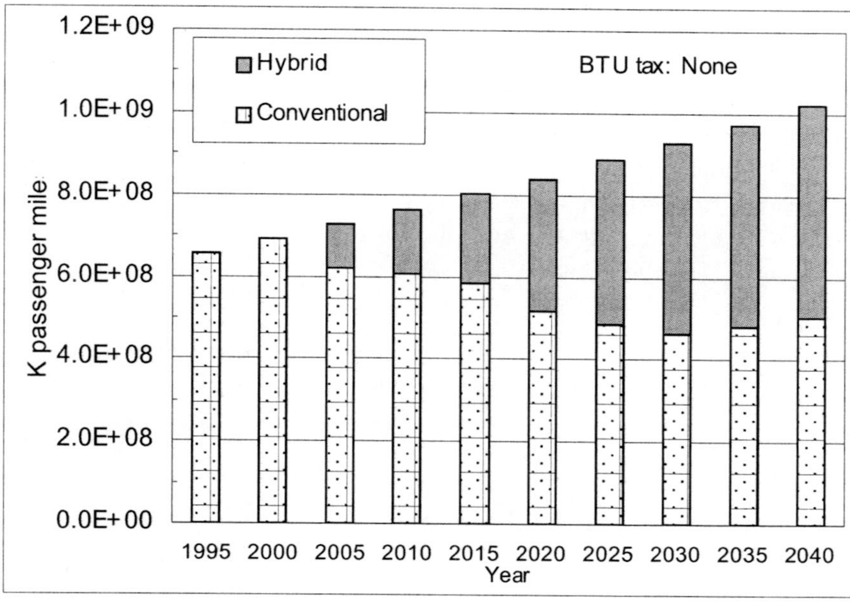

Figure 8: Passenger transportation using hybrid vehicles

Second, the changes in the passenger transportation sector resulting from introducing BTU taxes as well as hybrid vehicles are shown in Figure 9. A BTU tax of $4.5/mmBTU is charged on fossil fuels such as gas, petroleum, and coal. This results in an increase in the share of hybrid vehicles, which reaches 75 percent in 2040.

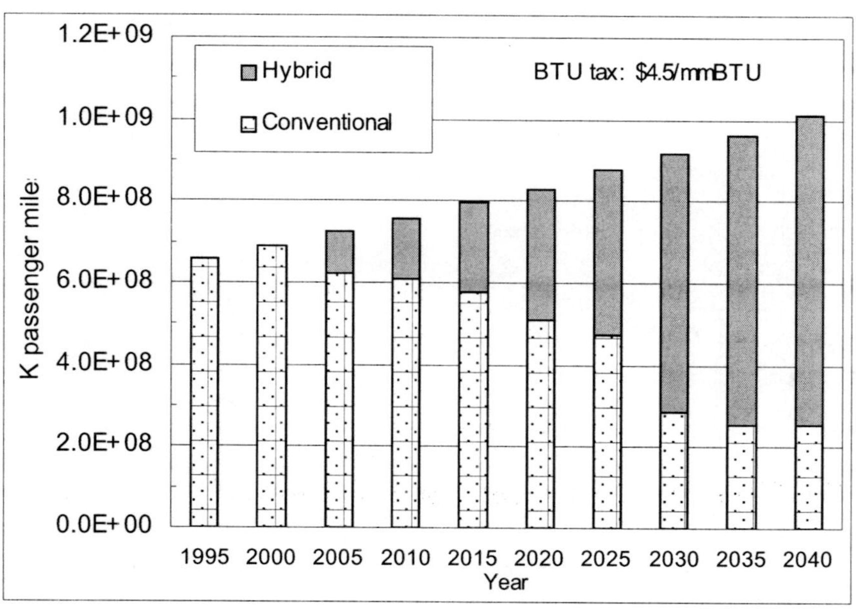

Figure 9: Passenger transportation using hybrid vehicles under the BTU taxes scenario

Third, the changes in energy consumption in the transportation sector under the BTU tax scenario are shown in Figure 10. There are four ways that the system has adjusted to lead to a reduction in carbon emissions:

1. There is a reduction in the use of conventional vehicles.
2. There is an increase in the use of hybrid vehicles. However, the use of these is relatively low because of higher fuel efficiency, which result in a lower I-O coefficient in the study.
3. There is an increase in the use of trucks.
4. There is a reduction of total energy use for transportation. This is mainly caused by the increase of hybrid vehicles.

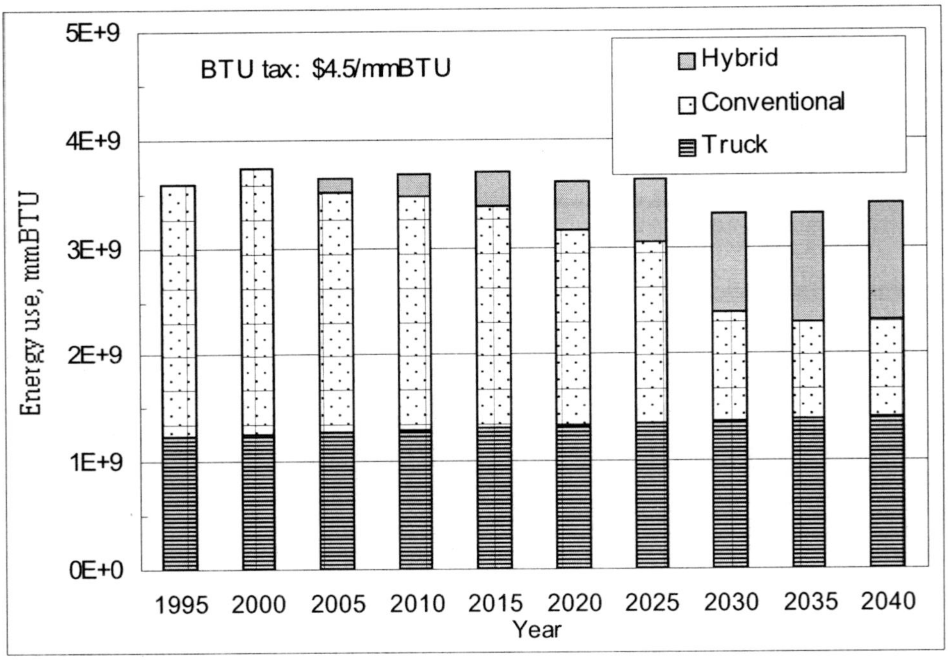

Figure 10: Total energy consumption in transportation sector

Total energy use in each resource sector in 2040 under the four scenarios is illustrated in Figure 11. We can see that the BTU tax reduces coal use the most. Under BTU taxation the cost of fossil fuels becomes higher, which orients the energy user toward more energy efficient choices. The BTU tax encourages the use of renewables, which enter the electricity market at competitive prices.

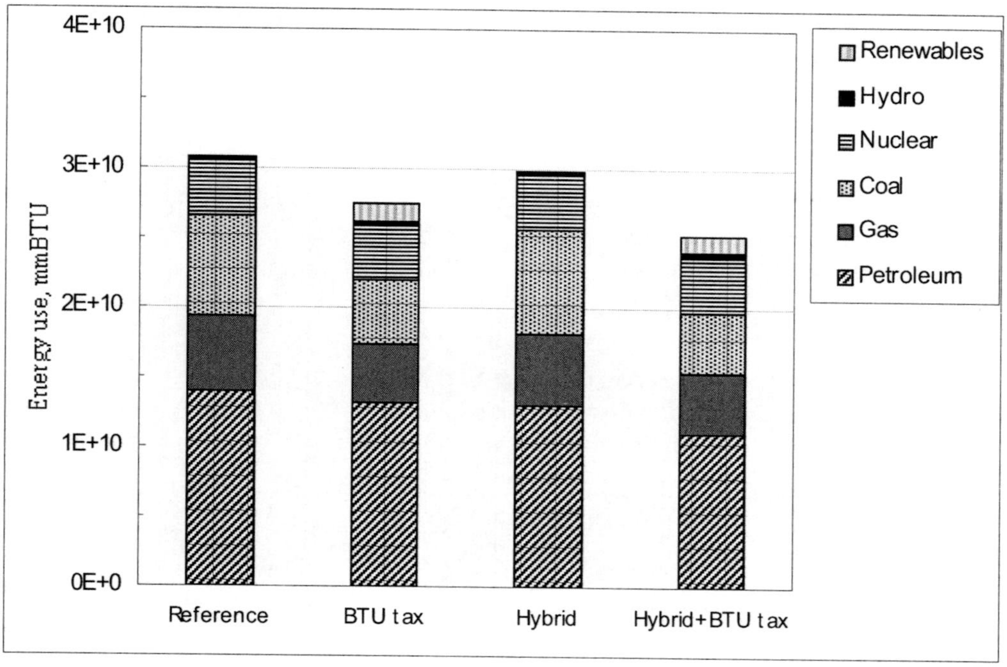

Figure 11: Total energy use in resource under the four scenarios in 2040

A summary of CO_2 emissions for the four scenarios is shown in Table 7. We can observe the following impacts of hybrid vehicles:

- Hybrid vehicles are effective to reduce carbon emissions from petroleum consumption for transportation use.
- With hybrid vehicles, CO_2 emissions rise to 350 mmTC by the year 2040 under the BTU tax scenario, producing a 122 mmTC reduction compared with the tax-free case.
- A BTU tax is effective to decrease carbon emissions from petroleum consumption for non-transportation use. In other words, carbon emissions from petroleum consumption for transportation use cannot be reduced even when BTU taxes are imposed.

Table 7: Summary of CO_2 emissions for the four scenarios in 2040

	Without hybrid vehicles		**With hybrid vehicles**	
	Reference	**BTU tax $4.5/mmBTU**	**Reference**	**BTU tax $4.5/mmBTU**
Petroleum	230	213	211	180
(Transportation)	(97)	(95)	(76)	(65)
(Non-transportation)	(133)	(118)	(135)	(115)
Gas	74	61	73	62
Coal	187	114	188	108
Total	491	388	472	350

Changes in the costs to consumers are seen in Table 8, which shows a summary of the costs and tax revenues for the four scenarios. The costs are the discounted present values of the total stream costs over the time horizon. In the hybrid vehicle cases, the capital costs are relatively higher than in the conventional vehicle cases because of the high capital costs of producing hybrid vehicles. In terms of fuel costs, we can see slightly less expensive costs using hybrid vehicles. Considering the total direct energy costs and the BTU tax revenue, the total cost to consumers is increased under the hybrid vehicles with BTU tax scenario. However, these differences in the cost to consumers are small, remaining within a ten-percent difference.

Table 8: Summary of costs and tax revenues for the four scenarios. The costs are the discounted present values of the total stream of costs over the time horizon. The discount rate was 5%. Unit: US$ billion.

	Without hybrid vehicles		With hybrid vehicles	
	Reference	BTU tax $4.5/mmBTU	Reference	BTU tax $4.5/mmBTU
Capital costs	2.12E+12	2.14E+12	2.31E+12	2.42E+12
Operating costs	5.27E+11	5.14E+11	5.18E+11	5.24E+11
Fuel costs	3.70E+11	3.57E+11	3.63E+11	3.42E+11
Total direct energy costs	3.02E+12	3.01E+12	3.19E+12	3.29E+12
BTU tax revenue	0	9.06E+10	0	8.47E+10
Total cost to consumers	3.02E+12	3.10E+12	3.19E+12	3.37E+12
Total energy consumption in 2040 (Quads)	30.8	27.4	29.9	25.2

Another consideration is the differences in total energy consumption for the four scenarios. The case involving hybrid vehicles and BTU taxes results in less energy consumption. In this case, 25.2 Quads represents an 18 percent saving in energy consumption compared with the reference case scenario.

4. Impacts of Nuclear Phase-Out on Energy Systems in Japan

4.1 Policy Scenarios Analyzed

Phasing-out of nuclear power is hypothesized to understand the role of nuclear energy for both strengthening energy security and reducing CO_2 emissions in Japan. In this investigation, a reference case analysis was first developed to model the expected emissions, resource use, and investments in energy technologies in the absence of policy actions. The reference case scenario means a business as usual condition. Under this scenario, it is assumed that there is no intervention in the energy system. In other words, reference case scenario is a base condition which result is already analyzed, evaluated and confirmed. The next was to model the effect of nuclear power phase-out and a carbon tax with a goal of a reduction of CO_2 by the year around 2040. A variety of approaches to reduce energy-related emissions of CO_2 and other greenhouse gases are possible. Under the term of The Kyoto Protocol, Japan would agree to reduce their carbon emissions by 6 % below 1990 levels, corresponding to a 22 - 25 % reduction from projections for 2010 (Environment Agency of Japan, 1998). This study focuses on a target rate of reductions of around 20 % of CO_2 by the year around 2040 since reaching the target in 2010 appears to be difficult and might not be realistic for Japan. We then determined the level of tax needed to reach the required level of CO_2 by the year around 2040. We evaluated the cost of that approach and the configuration of energy systems that results from it.

In the taxation case, the carbon tax was assumed to be imposed in increments, starting at zero in the first period and then rising by a fixed amount in each subsequent period until reaching the target amount of the tax. This approach avoids the sudden shock to the system and the stranded assets that can result from rapid market changes.

Reference Case Scenario

Nuclear boiler is assumed to be having constant generation quantity as it has in 1995 until 2040 considering government concern. Under this scenario, it is assumed that there is no intervention in the energy system, except that nuclear is subsidized as needed to maintain a 35% share of the electric market, the share in 1995. The trend of electric power generation under the reference case scenario is shown in Figure 3. Under this scenario, CO_2 emissions rise to 490 mmTC by the year 2040. The nuclear subsidy is our original term to remain the share of nuclear power boilers in the electricity sector at current output level. Without this subsidy, the price of nuclear power rises, resulting in nuclear's dying out in the electric market. The value of the subsidy is set at $9/mmBTU at most.

Nuclear Phase-Out Scenarios

In this case, we assume that the level of nuclear capacity will begin to phase-out in 2015 and will be reached at zero in 2035. Under this scenario, it is assumed that there is no intervention in the energy system. The CO_2 emissions rise to 530 mmTC by the year 2040.

Carbon Tax Scenarios

A carbon tax is expected to be the most efficient approach to reduce CO_2 emissions. It has been already implemented in Sweden, Finland, Norway, Denmark and Holland in range of $200/tonC (Sawa, 1997). This study evaluated several different tax rates to determine the rate that would achieve the target reduction by the year 2040. Both cases –with and without a carbon tax- were extrapolated to explore the difference in fuel prices. We found that a tax rate of $160/tonC produced an emission rate of 390 mmTC by the year 2040, achieving a reduction of 20 % in the emission rate.

Nuclear Phase-Out with Carbon Tax Scenarios

In the case of nuclear phase-out, at $205/tonC of carbon tax, CO_2 emissions rise to 390 mmTC by the year 2040. Again, the tax rate was gradually increased over time.

4.2 Discussion for Nuclear Energy

The discussion in this section highlights the changes in the energy system that result from reducing CO_2 emissions through carbon taxes, and the costs of implementing these changes.

Changes in the electricity sector in nuclear phase-out

Figure 12 illustrates the changes in the electric power generation under the nuclear phase-out case. The total energy consumption and the CO_2 emissions for four scenarios in the year 2040 are shown in Table 9. We can see three ways in which the system has adjusted to make up the nuclear boiler after its phasing out:

- The use of coal boiler and coal IGCC rise, and the total coal consumption rises by four times.
- The use of gas combined-cycles and gas boiler rise gradually, and the total gas consumption grows by three times.
- The renewables is not seen in the electricity market.
- There is a reduction of total energy use. This is mainly caused by the increase in efficiency in power generation shifting from nuclear boiler to gas boiler and gas combined-cycle.

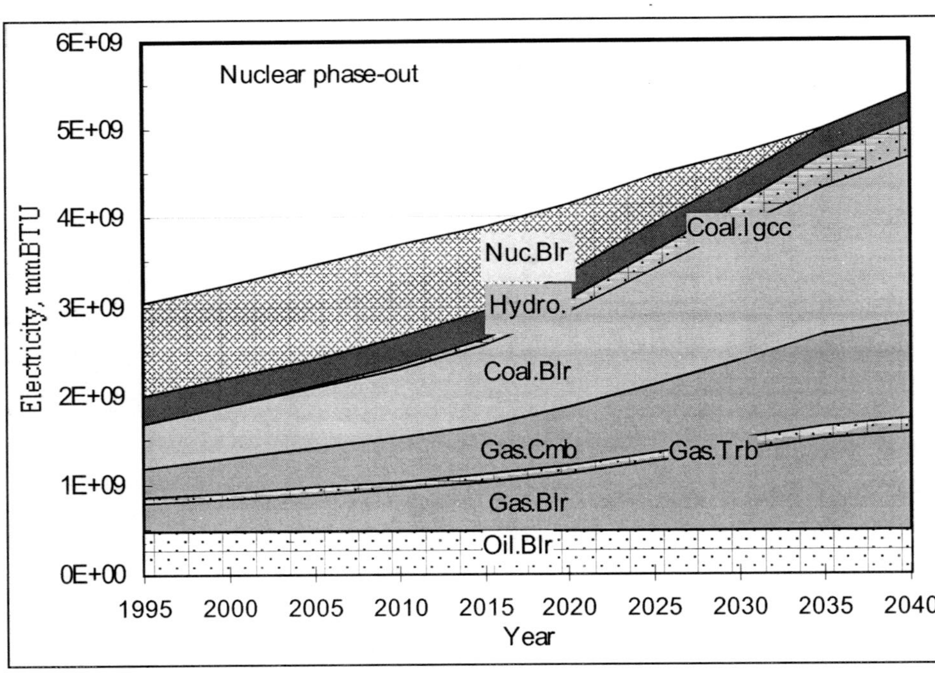

Figure 12: Electric power generation under the nuclear phase-out

Table 9: Summary of costs and tax revenues for the four scenarios. The costs are the discounted present values of the total stream of costs over the time horizon. The discount rate was 5%. Unit: US$ billion.

	Reference	Nuclear phase-out	Carbon tax $160/tonC	Nuclear phase-out & carbon tax $205/tonC
Capital costs	2.12E+12	2.10E+12	2.10E+12	2.09E+12
Operating costs	5.27E+11	5.15E+11	5.12E+11	5.03E+11
Fuel costs	3.70E+11	3.73E+11	4.31E+11	4.44E+11
Total direct energy costs	3.02E+12	2.99E+12	3.04E+12	3.04E+12
Carbon tax revenue	0	0	6.30E+10	7.97E+10
Total cost to consumers	3.02E+12	2.99E+12	3.11E+12	3.12E+12
CO_2 emission in 2040 (mmTC)	490	530	390	390
Total energy consumption in 2040 (Quads)	30.8	29.0	28.4	25.2

Figure 13 shows the changes in the electric power sector under nuclear phase-out with the carbon tax condition. The carbon tax is set at $205/tonC to reach the emission rate of 390 mmTC by the year 2040. We can see three ways in which the system has adjusted to lead the reduction of CO_2 emissions:

- The use of coal boiler falls, and IGCC's dying out in the electricity market.

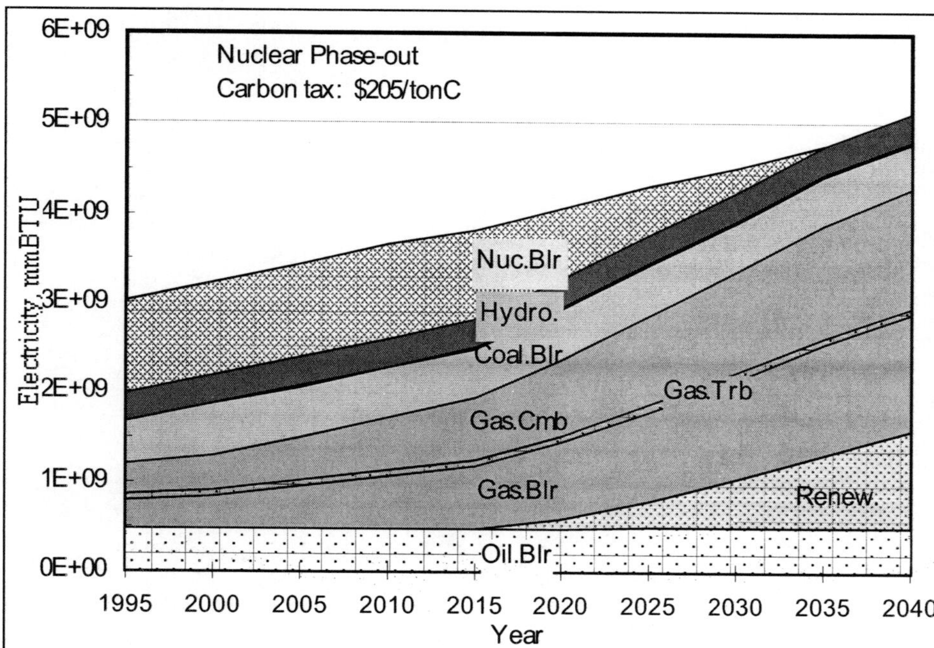

Figure 13: Electric power generation under the nuclear phase-out with carbon tax

- The use of gas combined-cycles and gas boiler rise, and the total gas consumption grows by four times.
- The use of renewables grows commencing the year 2015.
- There is a reduction of total energy use. This is mainly caused by the increase in price due to the taxes.

By comparing Figure 3, Figure 12 and Figure 13, we can get a sense of the difference in electricity supply. Nuclear power will be replaced by fossil fuels combustion such as coal boiler, gas boiler and gas combined-cycle. Because the price of electricity produced by coal is still cheaper than the price of gas fueled electricity, coal shows priority over gas fuel in case carbon tax is not charged. On the contrast, the price of coal rises more than the price of gas under the taxation case, resulting in the growth of gas power in the electricity supply (see Figure 14). The renewables are allowed to come into the market, when the price of renewable power becomes somewhat competitive under carbon taxes.

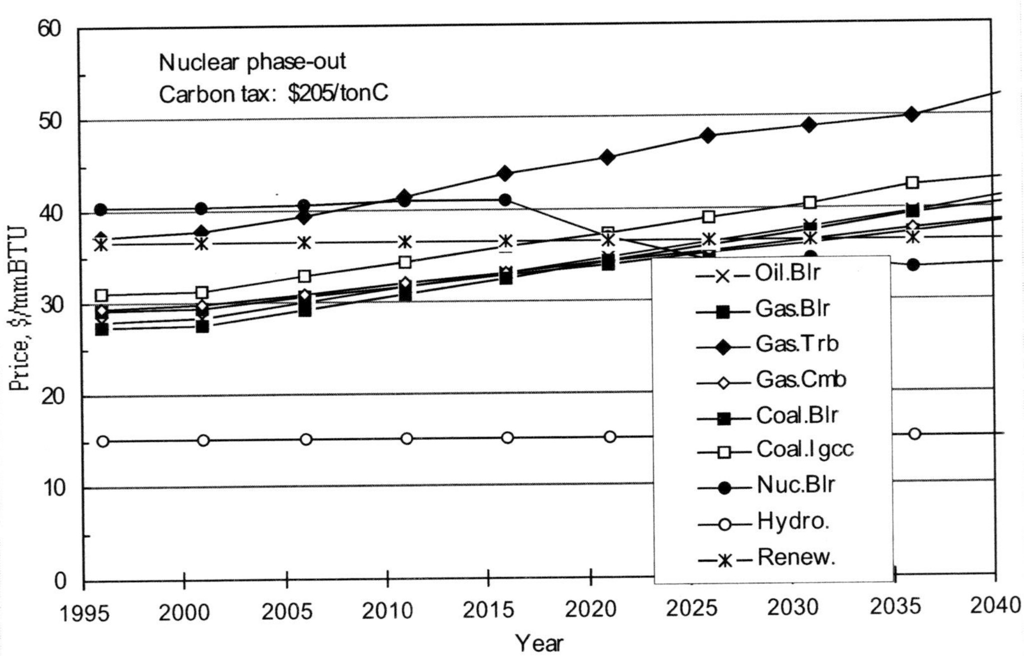

Figure 14: Changes in the price of electricity

Impacts of the Taxes with or without Nuclear Phase-Out

Figure 15 shows energy use in each resource sector under the four scenarios in the year 2040. We can observe the following impacts of a carbon tax:

- The carbon tax reduces the coal use the most.
- The carbon tax encourages the use of renewables.
- The CO_2 emissions rise to 530mmTC by the year 2040 under nuclear phase-out condition, which means 8 % increase compared with the reference case (see Table 9).

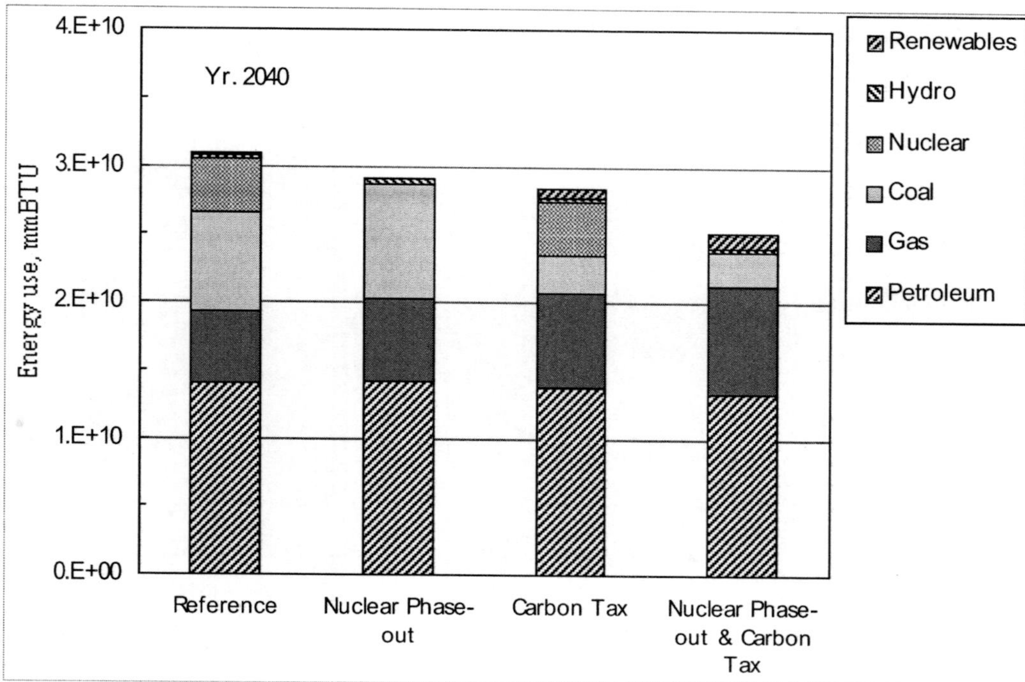

Figure 15: Total energy use in resource under the four scenarios in 2040

From the changes in the share of each resource, we can see the alternative scenarios without nuclear power. It looks possible to replace nuclear by fossil fuels as shown in Figure 6, however, the problem of local emissions and the fuel depletion are not considered adequately. Moreover, except for the slight reduction under the taxation case, the petroleum consumption remains stably at a constant level. Unlike the electricity sector, the transportation sector and some parts of the industrial sector have no alternatives to switch petroleum fuel easily. Renewables are another alternative which we can expect to produce electricity without emitting any environmental pollution. Besides its low intensity in energy conversion, renewable power requires optimum range of climate conditions to operate it. In the management of transmission grid, renewables cannot secure stable distribution to end-use customers.

Figure 16 shows the impacts of carbon taxes on CO_2 emissions. It is reasonable to suppose that a higher rate of carbon tax will be necessary when in nuclear phase-out to reach the same emission levels by the year 2040. This characteristic is mainly affected by the demand elasticities in each sector shown in Table 3. We carefully chose this value and confirmed its reliance in our previous study many times. However, there should be some changes in elasticities when carbon tax takes effect in the future. We believe that the trend in the reduction of CO_2 emissions by carbon taxation is reliable except the uncertainty of demand changes.

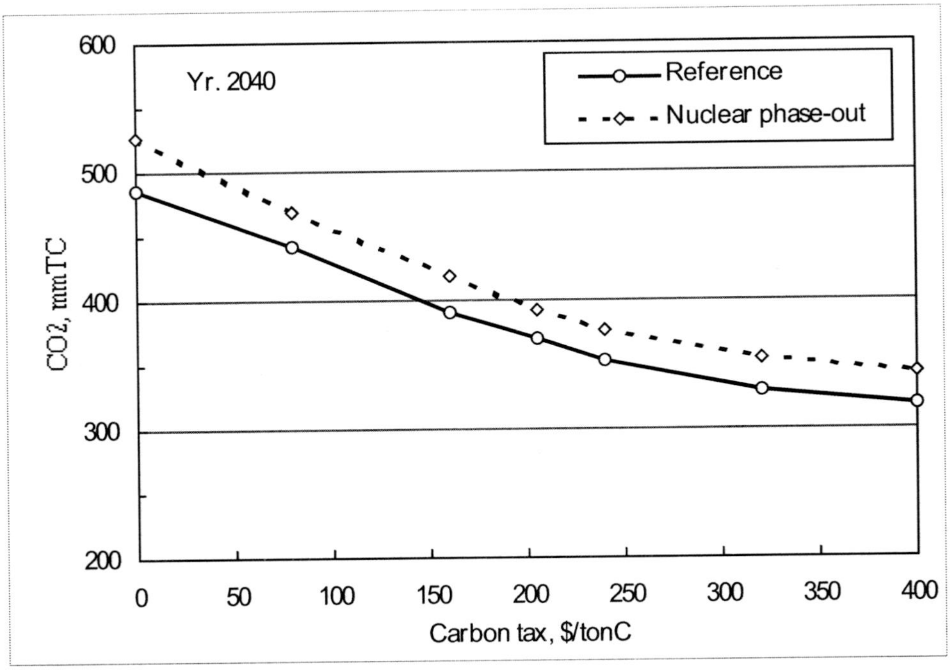

Figure 16: Impact of carbon taxes on CO_2 emissions

Changes in the Cost to Consumers

Table 9 shows a summary of the costs and tax revenues for the four scenarios. The costs are the discounted present value of the total stream of costs over the time horizon. The discount rate was 5 %. In the nuclear phase-out case, the capital costs are relatively lower than the reference case because of the absence of additional nuclear power stations. In terms of fuel costs, we can see slightly expensive cost under the nuclear phase-out case because of increased costs of fossil fuels.

Considering the total direct costs and the carbon tax revenues, the total cost to consumers will be increased under the nuclear phase-out with carbon tax scenario. However, these differences in the cost to consumers are small, remaining within a 5 % difference. Another consideration is the difference in total energy consumption for the four scenarios. The case involving nuclear phase-out and carbon taxes results in less energy consumption. In this case, 25.2 Quads represents an 18 % saving in energy consumption with the reference case scenario. The discussion of GDP, which could be analyzed by using macro economy model, is not included in this study.

5. ENERGY SYSTEMS IN KOREA

5.1 Policy Scenarios Analyzed

In the Republic of Korea, as the gross domestic product has increased rapidly, energy consumption also has increased at an annual rate of more than 10 percent since 1985 till the IMF crisis in 1997 (Table 10). Among the energy sources while the shares of coal and hydropower have decreased, the share of petroleum and liquefied natural gas have increased (Asia Development Bank, 1998, Korea Energy Economics Institute, 2000). Most of fossil energy is imported from abroad except small amount of domestic coal sources. The emissions of GHGs including CO_2 have increased in proportion to the energy consumption. The consumption of energy is closely related to economic growth and environmental problems. Therefore it is necessary to establish and implement energy policies for the balanced pursuit of economic development and environmental conservation.

Table 10: Energy statistics for two countries

		GDP, billion US$	Population, million	Primary energy consumption, Mtoe	CO_2, mmTC
1973	Japan	2,561	108	324	245
1973	Korea	89	34	21	18
1998	Japan	5,319	126	510	312
1998	Korea	516	46	163	100

Sources:

- OECD, International Energy Agency. (2000). *Energy prices and taxes*. Paris.
- OECD, International Energy Agency. (2000). *Energy balances of OECD* countries. Paris.
- OECD, International Energy Agency. (2000). *Energy statistics of OECD* countries. Paris.

Both two cases, carbon taxation case and the BTU taxation case, are analyzed by designing an energy-economic model for Korea. Necessary data are given by current references (OECD, 2000) and by some Korean colleagues (Shin, 2001). The results are shown in Figure 17. We can expect reductions of CO_2 emissions in the future by introducing carbon or BTU taxation into Korean economy. The carbon tax tends to suppress the consumption of carbon-heavy fuels such as coal, whereas the BTU tax appears to reduce the overall fossil fuels. In this sense, the BTU tax would be favorable for Korean energy security to reduce CO_2 emissions we well as keeping stable shares among energy sources.

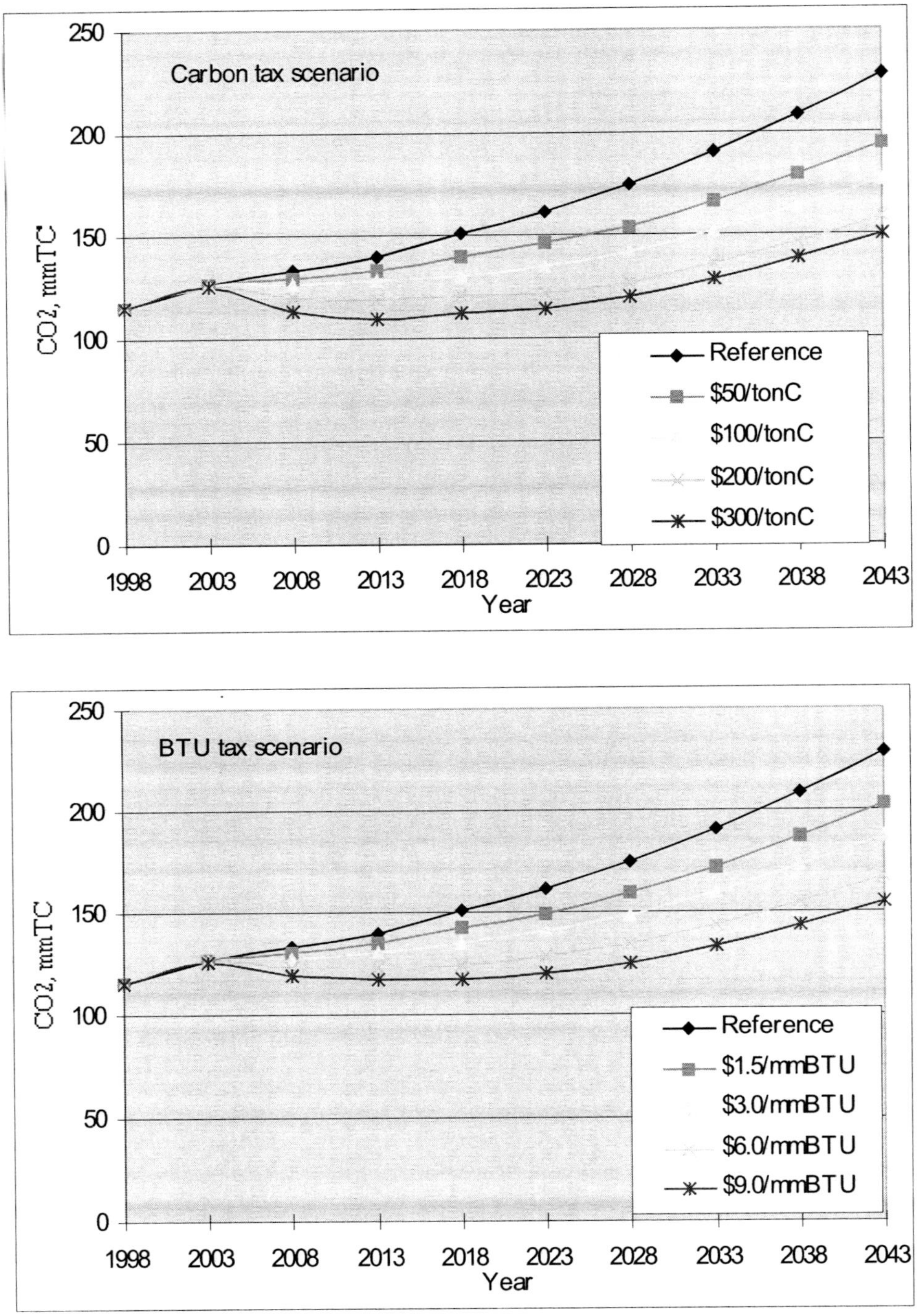

Figure 17: Carbon tax and BTU tax scenarios in Korea

5.2 Discussion for Energy Cooperation between Japan and Korea

There are some differences between Japan and Korea for current energy industries.

In order for competition to develop in electricity markets, monopolistic activities such as the operation of the transmission network need to be effectively separated from the potentially competitive activities such as power generation (OECD, International Energy Agency, 2001). The main objective of unbundling is to avoid discrimination in the competitive segments of the ESI (Electricity Supply Industry). On April 2, 2001, the Government of Korea unbundled former utility KEPCO (Korean Electric Power Cooperation) into new six power generation subsidiaries, and a power exchange. As the Korean electric power monopoly is ending, competition is being introduced in the process of restructuring. Customers will create a new order in the competitive electric power marketplace. Most of the functions and roles of the government will be replaced by market forces and behavior of customers which will be the most crucial factors in power markets.

On the contrast, Japan still looks to be studying what measures should be taken. The retail electrical market was liberalized in March 2000 for high-demand customers, which represented approximately 30 % of total electricity sales. This was the first attempt to catch up with the international standard of deregulation in fifty years' monopolistic history of Japanese ESI. As a result, companies from other industries have entered the power generation business and the practice of bidding for electrical power has expanded. Growth in power consumption, which has steadily increased since the utility's founding, is distinctly slowing amid weak economic expansion and advancements in energy conservation. However, the rest of 70 % customers are still put in the classical situation where neither choice nor competition is allowed for both customers and utilities. The price of electricity per kWh for residential use is 28 JPY (23 US cents), which shows the highest fare in OECD countries.

Along with the market changes toward deregulation and competition, we can see possible approaches for energy cooperation between Japan and Korea. First, there can be competitive pools or power exchanges such as NordPool between two countries. NordPool is a voluntary electricity exchange open to traders from Norway, Sweden, Finland and parts of Denmark. As of 1997, over 40 % of electricity trade in that area was handled by the pool. The geographical location between Japan and Korea is similar to that of Scandinavian countries, and the electric utilities of both countries have point of similarity in generation, transmission and distribution systems. The new pool market can give a chance to operate the ESI in efficient way as well as empowering the end user, and meeting security of supply, environmental and social goals.

Second, Japanese government officials and Exxon Mobil Corporation executives are talking up a 2,100 km pipeline that would carry natural gas from the Russian far East to either Tokyo or Niigata, halfway down the Japanese archipelago (Business Week, 2001). In Korean peninsula, the domestic gas pipeline has been built between Seoul and Pusan, and its extension from China to Seoul is under discussion by foreign investors. Provided that a pipeline between Korea and Japan is designed, this network could secure natural gas supply for both countries. The current LNG price for industry in Japan is \$96.45 in 2000 which is unnaturally high compared with \$41.95 in UK and \$26.50 in the US (OECD, International Energy Agency, 2000). Regional integration of gas market and the electricity network will make an important role for two countries.

CONCLUSIONS

Both the carbon tax and the energy tax can be effective in reducing nation's carbon emissions. The total cost to the consumer will be approximately the same under either scenario — although the energy tax will result in larger tax revenues. However the carbon tax tends to eliminate coal as an energy resource for Japan. Since Japan must import nearly all of it fossil fuels, narrowing its mix would leave Japan more vulnerable to price variations in the international petroleum and natural gas markets and would reduce their bargaining power in negotiations. Thus, the energy tax would be a more stable approach for Japan.

The transportation sector uses one-fourth of total primary energy. Because there are no alternative fuels for transportation except petroleum, the effects of carbon or energy taxes are relatively small. Hybrid vehicles, which have excellent fuel efficiency, could be one of the breakthrough technologies to meet the environmental constraints. Either carbon or energy taxes would encourage the penetration of hybrids.

In the long run, renewable resources may contribute more to carbon reductions than envisioned here. In our study renewables are only included in the electric sector, as renewable technologies in other sectors are not widely available. If renewable technologies could be used in the transportation sector (e.g. through electric or hydrogen vehicles fueled by renewable generation), petroleum consumption and CO_2 emissions would be drastically reduced.

Moreover, there are uncertainties about the future capacity of nuclear power. MITI recently announced that nuclear power stations could be technically operable for sixty years under strict maintenance conditions. This suggests that the Government may change its nuclear policy from building new stations to extending the life of existing stations. Over the long term this could result in a decrease in the share of nuclear capacity. Since our study is based on the assumptions that nuclear will supply 35 percent of electricity to the year 2040, a decrease in nuclear power will affect the capacities in electric power generation. Nuclear' s role in abating carbon emission should be re-evaluated as the government's policy becomes clearer.

Based on these analyses, we can draw the following conclusions:

- Achieving carbon reductions *with the set of technologies considered here* will mean reduction in energy use as well as shifts to lower carbon fuels. Introducing other technologies, such as a greater reliance on renewables or hybrid vehicles, might ameliorate the projected reduction in energy use.
- The total cost in terms of supplying energy will be similar for all three cases. However, the carbon reduction cases imply reduction in energy use. This will probably be achieved through more efficient end use process or other shifts which may require additional capital investment (though it is not clear that higher capital investment is inevitable)
- The carbon taxes have their advantages and disadvantages. The carbon tax is efficient in that smaller tax revenues are raised and the energy use reduction is smaller. However, the energy tax maintains a diversity of energy resources which is an important consideration for Japan's energy security.

Second, the main impact of hybrid vehicles is a decrease in petroleum consumption in the transportation sector (Nakata, 2000). As hybrid vehicles are introduced into the passenger transportation market, the share of conventional vehicles decreases proportionally, leading to a decrease in CO_2 emissions.

BTU taxation has a positive impact on reducing the consumption of fossil fuels such as coal, gas and petroleum for non-transportation purposes. As there are no alternative choices in the transportation sector, BTU taxation has no effect on petroleum consumption for transportation. However, when hybrid vehicles become available, higher fuel prices promote a shift by consumers from conventional vehicles to hybrid vehicles, which results in reducing carbon emissions drastically.

From the viewpoint of financial parameters, higher dependence on hybrid vehicles involves relatively high capital costs. The total cost in terms of supplying energy will be similar for all four cases. However, the carbon reduction cases imply a reduction in energy use. This would probably be achieved through more efficient end-use processes or other shifts which would likely require additional capital investment.

Third, the main impact of nuclear phase-out is an increase in fossil fuels consumption in the electricity sector (Nakata, 2002). As nuclear power begins phasing out, the output of coal and gas fired power generation grows proportionally, leading to an increase in CO_2 emissions.

A carbon tax has a positive impact on reducing the consumption of fossil fuels such as coal and gas. The carbon tax will raise the price of high-carbon fuels, such as coal, and lead to greater reliance on low-carbon fuels. Even for clean coal technologies such as coal IGCC does not penetrate the electricity market. The carbon tax tends to eliminate coal as an energy source for Japan.

From the viewpoint of financial parameters, higher carbon tax will be necessary under nuclear phase-out scenario to achieve the same target of CO_2 emissions. However, carbon tax scenario tends to be relying on gas resource increasingly. Taking into consideration of energy security issues, BTU tax scenario (Nakata, 2001) would be more stable solution for Japan where all the fossil resources are imported.

Another consideration is the use of renewable resources and the technologies for their conversion into energy. In our study, renewables are only seen in the electricity sector, as renewable technology in other sectors is not yet on the horizon. If future technology were to enable the use of renewable resources in the transportation sector, decreased petroleum consumption could contribute lead to drastic reductions in CO_2 emissions.

Based on these analyses, we can draw the following conclusions:

- Nuclear power phase-out will mean shifts from nuclear to fossil fuels such as coal and gas, resulting in increase in CO_2 emissions in the electricity sector.
- There is a dramatic difference in the carbon tax between reference case and nuclear phase-out case. To achieve the same level of CO_2 emissions without nuclear power in the future, it will be necessary to charge a 30 % higher carbon tax than in a similar scenario with nuclear power. However, the presented results are derived from a fixed set of initial condition such as price elasticity in energy demand. It is therefore necessary to note some changes in real energy supply in the future caused by unexpected matters.

Finally, the regional pool market of electricity in Japan and Korea are discussed. The extension of gas pipelines is also suggested to secure energy supply and to meet environmental and social goals in efficient way. The concept of energy grid between two countries will make an important role both in strengthening energy security with economical way and in building a more market-oriented economy, resulting in stabilized political situation in East Asia.

REFERENCES

Asia Development Bank, 1998. Asia least-cost Greenhouse gas abatement strategy, Republic of Korea. Manila, Philippines.

Allenby, BR., 2000. Environmental security: Concept and implementation. International Political Science Review, **21**(1), 5-21.

Andersson, B. and Haden, E., 1997. Power production and the price of electricity: An analysis of a phase-out of Swedish nuclear power. Energy Policy, **25**(13), 1051-1064.

Banks, F., 2000. Oil, electricity deregulation, and nuclear energy. Petromin, March 24-27.

Bodde, D., 1998. Strategic thinking about nuclear energy: Implications of the emerging market structure in electric generation. Energy Policy **26**(12), 957-962.

Does Japan really need a $16 billion gas pipeline? 2001. Business Week, October 22, 37.

Campbell, C. and Laherrere, J., 1998. The end of cheap oil. Scientific American, March 78-83.

Chung, W., Wu Y. J. and Fuller, J. D., 1997. Dynamic energy and environmental equilibrium model for the assessment of CO_2 emission control in Canada and the USA, Energy Economics **19**, 123-124.

Doucet, G., 1999. The inevitability of nuclear power generation. Proceedings of the Uranium Institute 24th Annual Symposium, 1-6.

Edmonds, J.A., Pitcher, H.M., Barns, D., Baron, R. and Wise, M.A, 1995. Modeling future greenhouse gas emissions: The second generation model description. In modeling global change, ed. Klein L. R. and Lo, F., pp. 295-362. United Nations University Press Tokyo.

The Energy Data and Modeling Center, 1998. EDMC Handbook of Energy & Economic Statistics in Japan, Tokyo.

Environment Agency of Japan, 1998. White Paper on Environment, Tokyo.

The Federation of Electric Power Companies, 1998. Handbook of Electric Power Industry, Tokyo.

Gan L., 1998. Energy development and environmental constraints in China. Energy Policy **26**(2) 119-128.

Goto, N., 1995. Macroeconomic and sectoral impacts of carbon taxation. Energy Economics **17**(4), 277-292.

Goto, N. and Sawa, T., 1993. An analysis of the macroeconomics costs of various CO_2 emission control policies in Japan. The Energy Journal **14**(1), 83-110.

Haas, R. and Auer, H., 2001. How to ensure effective competition in western European electricity markets. International Association for Energy Economics Newsletter, Third quarter, 16-20.

Hoster, F., 1998. Impact of a nuclear phase-out in Germany: Results from a simulation model of the European Power Systems. Energy Policy, **26**(6), 507-518.

MITI's new ten-year plan calls for development of 11.3million kW nuclear power to raise its generating capacity to 56.1 million kW by end of FY 2007, 1998. JPET, May; 2-9.

Kagramanian, V., Kononov, S. and Rogner, H., 2000. Climate change driving forces, Nuclear energy & the latest IPCC emission scenarios. IAEA Bulletin, **42**(2), 31-35.

Kaya, Y., 1997. Japanese strategy for mitigating global warming. Energy Conversion and Management, **38**, 19-23.

Kibune, H. and Kudo, H., 1996. Structural changes in Japan's economy and society and outlook for long-term energy supply and demand. Energy Policy **24** (12), 1119-1125.

Kiso, A., 1998. Trends of nuclear power development in Asia. Energy Policy **26**(7), 577-582.

Korea Energy Economics Institute, 2000. Yearbook of energy statistics, Seoul.

Lamont, A., 1994. User's Guide to the META•Net Economic Modeling System Version 1.2. UCRL-ID-122511, Lawrence Livermore National Laboratory, Livermore, California.

Lazarus, M., Heaps, C. and Raskin, P., 1997. LEAP Long-range energy alternatives planning system. Stockholm Environmental Institute, Boston, Massachusetts.

Lee, H., Oliveira-Martins, J. and Mensbrugghe, D., 1994. The OECD green model: An updated overviews. Technical paper No. 97, OECD, Paris.

Lydick, J., Morris, S.C., Lee, J. and Goldstein, G., 1990. Demo MARKAL abbreviated version of the US MARKAL energy systems model, created for demonstration: abilities, limitations, and demonstration, BNL-47782. Brookhaven National Laboratory, Upton, New York.

Matthews, J., 1998. Challenges for nuclear power in Japan, Korea and Greater China at the turn of the century. Nuclear Energy **37** (4), 233-239.

MITI, Agency of Natural Resources and Energy, 1998. Interim Report on General Energy Supply and Demand, Tokyo.

Nagata Y. Personal communication. June 21, 2000.

Nakata, T., 2000. Analysis of the impact of hybrid vehicles on energy systems in Japan. Transportation Research, Part D **5**, 373-383.

Nakata, T. and Lamont, A., 2001. Analysis of the impacts of carbon taxes on energy systems in Japan. Energy Policy, **29**(2), 159-166.

Nakata, T., 2002. Analysis of the impacts of nuclear phase-out on energy systems in Japan. Energy, printing.

Nordhaus, W., 1997. The Swedish nuclear dilemma: Energy and the environment. Resources for the Future, Washington, D.C.

OECD, 2000. OECD economic surveys, Korea. Paris.

OECD, International Energy Agency, 2000. Energy prices and taxes. Paris.

OECD, International Energy Agency, 2000. Energy balances of OECD countries. Paris.

OECD, International Energy Agency, 2000. Energy statistics of OECD countries. Paris.

OECD, International Energy Agency, 2001. Competition in electricity markets. Paris.

Richels, R. and Sturm, P., 1996. The costs of CO_2 emission reductions. Energy Policy **24** (10/11), 875-887.

Romerio, F., 1998. The risks of the nuclear policies. Energy Policy **26**(3), 239-246.

Sato, O, Tatematsu, K. and Hasegawa, T., 1998. Reducing future CO_2 emissions – The role of nuclear energy. Progress in Nuclear Energy **32**(3/4), 323-330.

Sawa, T., 1997. Chikyuu Ondanka Wo Fusegu. Iwanami Shoten Publishers, Tokyo.

Shin, E., 2001. Personal communication. August 21, 2001.

Schultz, P. A. and Kasting, J. F., 1997. Optimal reductions in CO_2 emissions, Energy Policy **25** (5), 491-500.

US Department of Energy, Energy Information Administration, (1998a). Impacts of the Kyoto Protocol on U.S. energy markets and economic activity. Washington D.C.

US Department of Energy, Energy Information Administration, (1998b). Assumptions to the Annual Energy Outlook, Washington D.C.

Welsch, H., 1998. Coal subsidization and nuclear phase-out in a general equilibrium model for Germany. Energy Economics, **20**, 203-222.

Welsch, H. and Ochsen, C., 2001. Dismantling of nuclear power in Germany: sectoral and macroeconomic effects. Energy Policy, **29**, 279-289.

Yamaji, K. Matsuhashi, R. Nagata, Y. and Kaya, Y., 1993. A study on economic measures for CO_2 reduction in Japan. Energy Policy **21** (2), 123-132.

Zongxin, W. and Siddiqi, T., 1995. The role of nuclear energy in reducing the environmental impacts of China's energy use. Energy **20**(8), 777-783.

Chapter 4

ECONOMIC EFFECTS OF NUCLEAR PHASE-OUT IN GERMANY

Heinz Welsch and Carsten Ochsen
Department of Economics
University of Oldenburg
26111 Oldenburg
Germany

ABSTRACT

The German government and the German electricity industry have recently agreed on phasing out nuclear power. The agreement entails cutting the lifetime of nuclear power stations to 32 years. The paper examines the economic and environmental effects of nuclear phase-out, utilizing a dynamic computable general equilibrium model which encompasses a detailed representation of the electricity supply industry. The nuclear phase-out entails an increase of the electricity price by up to 8 percent and an ensuing fall in electricity demand by up to 3.7 percent. The remaining supply gap is filled by additions to fossil-fuelled capacity. The macroeconomic effects display considerable dynamics which reflect variations over time of the relative weight of expansionary capacity-replacement impulses and contractionary electricity price effects. While the macroeconomic effects of nuclear phase-out are rather moderate on average, carbon emissions get significantly enhanced.

1. INTRODUCTION

The use of nuclear power has been a highly controversial issue in German energy policy for considerable time. While additions to nuclear capacity were not contemplated by any of the parties involved, there was considerable controversy as to how long existing capacities should be used. The electricity industry insisted on a lifetime of 40 years for existing nuclear plants, implying that the last nuclear station would be decommissioned in 2029. The Green Party initially demanded that nuclear stations should be shut down after 25 years of utilisation, but has later agreed to the official view of the German government, which stipulated a 30 year lifetime.

In June 2000, the government and the industry agreed on cutting the lifetime of nuclear power stations to 32 years. Following this decision, a number of questions arise:

- Which electricity generation technologies (energy carriers) will fill the supply gap?
- What will be the effect on electricity prices, electricity consumption, and (possibly) electricity imports?
- What will be the effect on the non-energy industries and the economy at large?
- What environmental effects will arise, especially in terms of CO_2 emissions?

This paper provides an assessment of these issues with the help of a computable general equilibrium (CGE) model which encompasses an appropriately detailed representation of the electricity supply industry. In comparison with previous studies on (variants of) a nuclear phase-out in Germany, the contribution of the present paper lies in explicitly addressing the phase-out scenario actually agreed, and in using a methodology that is able to analyse both the electricity-related and the general-economic issues in a consistent way. By contrast, earlier papers have studied ad-hoc phase-out scenarios or/and have employed partial-analytical techniques.

Hoster (1998) compares two ad-hoc scenarios: a short-term phase-out to be completed by 2005 and a scenario of constant nuclear capacity, using a linear-programming model of the electricity industry with an exogenous demand profile. The same issue is addressed in Hoster, Welsch and Böhringer (1997, Chapter 6), using a model in which the linear-programming electricity model is embedded in a general equilibrium model, thus endogenising electricity demand. A more long-term issue, namely a dismantling by 10 percent annually compared to a constant-capacity scenario, is analysed by Welsch (1998) with the help of a CGE model which includes a special treatment of the electricity industry.

The problem of these three studies is that constant nuclear capacity is not a realistic base case. This point is taken up by Pfaffenberger and Gerdey (1998) and Pfaffenberger (1999). They use the 40-years-lifetime scenario as their base case and analyse the effects of a short-term phase out to be accomplished by 2005. Their method is to employ a linear-programming model of the electricity industry to compute required capacity additions in terms of hard coal and natural gas (keeping other energy carriers and the required electricity output fixed). The resulting investment requirements and electricity costs are fed into a macro-sectoral simulation model to compute the effects on the overall economy.

The current paper can be regarded as a follow-up to Welsch (1998) as far as methodology is concerned. It differs from that paper by addressing more realistic nuclear scenarios and using updated projections of fuel prices and the use of renewable energy. The paper is organised as follows. Section 2 describes the specification and parameterisation of the model. In section 3 our assumptions and scenarios are presented. Section 4 gives the results of our simulation analysis, and section 5 concludes.

2. Specification and Parameterisation of the Model

The nucleus of our methodological approach is a model of electricity generation and electricity costs, differentiated by type of fuel, which is embedded within a general equilibrium framework.

2.1 The General Equilibrium Model

This section gives a brief, non-technical description of LEAN-TCM, a two-region dynamic general equilibrium model of the European Community. A more detailed and technical statement of the model can be found in Welsch and Hoster (1995).

The regions distinguished in LEAN-TCM are West Germany (GER) and the rest of the European Community, excluding Greece and Luxembourg[1], (EC9). Modelling Germany in the context of the European Community is important because the German energy policy debate is characterised by a strong emphasis on international cost competitiveness, especially with respect to the European Internal Market.

Within each region there are 13 sectors, five of which refer to energy carriers: hard coal, brown coal, petroleum, gas and electricity. The other sectors are agriculture, intermediate goods, equipment, consumption goods, construction, transportation, services, and government. The model is recursively dynamic, with a time horizon to 2030.

As is common for computable general equilibrium models, foreign trade modelling follows the approach of Armington (1969), wherein goods of the same kind but different place of origin are treated as imperfect substitutes. The imperfect substitutability of goods of different origin is captured by constant elasticity of substitution (CES) aggregator functions. Rather than considering bilateral trade relations, foreign trade is modelled by means of a world trade pool, in which the world trade volume is the basic activity variable[2]. The world trade volume of each good is the sum across GER, EC9 and the rest of the world (ROW) of import demands.

Since German and EC9 imports are endogenous, a considerable fraction of the world trade volume, to be allocated to exports from GER, EC9 and ROW, is endogenous. The allocation of the world trade volume works in two stages. First, it is allocated to exports from ROW and exports from EC, utilising CES demand functions. Total EC exports are then subdivided among German exports and EC9 exports. The elasticity of substitution between German exports and EC9 exports is typically larger than that between EC exports and ROW exports.

Import demand is modelled in a similar way. Within each region, the overall amount of each good (being the sum of intermediate and final demand) is allocated to imports and domestic production by means of CES demand functions. Only for hard coal, domestically produced and imported quantities are considered to be homogeneous goods; coal imports are

[1] Greece and Luxembourg were not included due to poor data. The term EC refers to the member states as of 1993.
[2] This approach is motivated by the intention to disaggregate EC9 in future model versions.

determined as a residual from overall demand and (exogenous) domestic production (see below).[3]

The foreign trade model is completed by a simple mechanism for exchange rate determination. The EC is treated as an exchange rate union vis-a-vis ROW. Thus, two currencies may be said to exist in the model. The exchange rate between these currencies is assumed to react to changes in the EC's balance of current account vis-a-vis ROW. More specifically, it is assumed that the price of foreign exchange remains constant whenever the current account balance is in equilibrium. A negative balance represents excess demand for foreign exchange and brings about an increase in the exchange rate, while a positive balance works in the opposite direction. The parameter that governs this mechanism is the elasticity of the exchange rate with respect to the ratio of nominal EC imports to nominal EC exports (including trade in services).

As concerns primary inputs, labour is assumed to be immobile across borders and completely mobile across sectors. Aggregate labour supply within each country is described by a dynamic wage equation, according to which the current wage equals the wage of the previous period times the increase in labour productivity and the price level, modified by the ratio of actual employment to 'normal' employment (Phillips curve). Employment in each country is the sum of labour demand across sectors. For capital, free mobility is assumed. Then, there is a uniform (real) interest rate. It is determined by the consolidated flow of funds constraint of EC vis-a-vis ROW. Thus, the interest rate serves as the closure variable of the model.

Considering the final demand categories, we first note that the consumption expenditures of the representative household are a fraction of available labour and capital income. The savings ratio is assumed to depend on the interest rate, with a constant elasticity[4]. Total consumption expenditures are then allocated to consumption of the different goods, utilising the Linear Expenditure System.

Nominal macroeconomic investment is the sum of the sectors' investment in value terms. The price of investment goods is sector specific, since each sector's capital good is characterised by its specific composition in terms of sectors of origin. A sector's real investment is the difference between the capital stock considered optimal for the next period and that part of the current capital stock that will still be in operation in the next period. Investment demand for a sector's goods is the sum of sectoral investment requirements, weighted by the (constant) coefficients of the capital composition matrix.

Factor demand is derived from a five-stage nested production function for each sector, except electricity.[5] At the top level, output is linked to an aggregate of energy, capital and labour (EKL) and to the various intermediate inputs via constant input-output coefficients. The EKL aggregate is further broken down into labour and an energy-capital aggregate. This choice of disaggregation reflects our interpretation of capital as a collection of facilities for using energy. While the substitution possibilities *within* the energy-capital aggregate are

[3] Note that for the other energy carriers it is reasonable to assume imperfect substitutability of imports and domestic sources. This is to reflect various circumstances: The unimportance of international trade in brown coal, the inelastic (or non-existing) domestic supply of petroleum and gas, and institutional restrictions to trade in electricity. These circumstances imply a low elasticity of substitution between imports and domestic supply of these energy carriers.

[4] This encompasses the frequently-considered special case of a constant savings ratio.

[5] The electricity supply industry is treated differently, see below.

small, they are larger *between* energy-capital and labour[6]. Next, energy-capital is separated into capital and energy. Energy, in turn, is an aggregate of fossil energy and gas. Typically, the elasticity of substitution among fossil fuels is larger than that between fossil fuels and electricity. The latter, in turn, is larger than that between energy and capital[7].

Factor demand is derived from profit maximisation subject to the production structure just outlined. At the top level of the production process inputs are related to output via fixed coefficients. At the subsequent levels there are CES demand functions similar to those used in the foreign trade model. The sectoral capital stock in operation in any period is the capital stock considered optimal in the previous period. Thus, capital is a quasi-fixed factor. Energy, being a variable factor, adjusts to the predetermined capital stock.

The supply side is represented by the price model. The exogenous driving forces of the price model are the export prices of the rest of the world, expressed in the ROW currency, and the carbon/energy tax. Together with the wage rate and the interest rate they determine the product prices via price aggregator functions dual to the quantity aggregators (production functions) discussed above.

2.2 The Electricity Supply Model

In contrast to standard multi-sector general equilibrium models particular care has been devoted to modelling the electricity supply industry, though no attempt was made at including a detailed model of technology choice.

Electricity supply is differentiated by type of fuel input. In addition to nuclear and to renewable electricity, four types of fossil fuel are distinguished: hard coal, brown coal, oil, and gas. Differentiating electricity generation by type of fuel is mainly motivated by a desire to take into account differences in capital intensity. Both electricity output and fuel input are measured in thermal units. This allows explicit consideration of conversion efficiencies, which is not normally possible in models measuring output and input in money units (at constant prices).[8] To assure consistency of the electricity supply model with the overall general equilibrium model, fixed converters between the respective units of measurement are applied.

The structure of the electricity supply model is as follows. Total electricity (EL) is the sum of renewable electricity (EL_R), nuclear electricity (EL_N), and an aggregate of fossil-fuel based electricity (EL_F). Fossil-based electricity is broken down according to the fuel used (hard coal, brown coal, oil, gas). Each type of electricity is taken to be an aggregate of fuel input and capital input. Inputs of labour and intermediates are not differentiated by fuel; they are linked to overall electricity through fixed input-output coefficients.

Nuclear electricity and renewable electricity are exogenous variables. Thus, fossil-fuel electricity is obtained from total electricity demand (which is determined in the general equilibrium model) as a residual:

$$(1)\ EL_F = EL - EL_N - EL_R.$$

---

[6] For an extensive discussion in conceptual and empirical terms see Burniaux et al. (1992).

[7] For an overview of estimates see again Burniaux et al. (1992).

In contrast to the separation of total electricity into fossil-based and non-fossil-based electricity, the shares of the various fossil fuels in the fossil electricity aggregate are derived from cost minimisation subject to a fuel preference function. Fuel preferences capture the incomplete substitutability of fuels with respect to the load profile of electricity demand as well as considerations relating to the security of fuel supply.

Assuming that fuel preferences can be described by means of a CES function, the cost minimising quantity of electricity based on fuel *i* is as follows:

$$(2)\quad EL_i = \lambda d_i^{\sigma}\left(\frac{PEL_F}{PEL_i}\right)^{\sigma} EL_F$$

where σ =elasticity of substitution, d_i = distribution parameter, PEL_i = price of electricity from fuel i, PEL_F = price aggregate of fossil-based electricity, λ = scaling parameter (adjusted such that $\Sigma_i EL_i = EL_F$).

Each type of fossil fuel electricity output is taken to be an aggregate of fuel and capital input[9]. The production function (aggregator function) is supposed to have the following form:

$$(3)\quad EL_i = a_i\left(FEL_i - EL_i\right)^{b_i}\left(K_i^{'}\right)^{1-b_i},$$

where FEL_i = input of fuel i, $K_i' = \alpha_i K_i$= capital input 'in effective units'. Note that α_i is an index of the efficiency of existing capital, whereas K_i denotes the pure quantity. The efficiency factor α_i increases over time, depending on the exogenously growing efficiency of new-vintage capital and the endogenous share of new-vintage capital in the overall stock.

The specification (3) has been chosen because it exhibits several desired properties. Especially, it implies that small improvements of the conversion efficiency (heat rate) require only small increases in capital, while for the conversion efficiency to approach its limit ($EL_i/FEL_i \rightarrow 1$) an infinite amount of capital would be required.[10] In addition, it can be shown that the elasticity of substitution between fuel and capital is between zero and unity, approaching zero as the conversion efficiency increases. These are plausible qualitative features of aggregate electricity generation which are lacking in standard neoclassical approaches.

Treating capital as a quasi-fixed factor, the production function can be solved for fuel demand as a function of output and the existing stock of capital (capacity):

---

[8] If quantities are expressed in terms of monetary units at constant prices, electricity output is larger than fuel input, and the conversion efficiency (heat rate) is not an applicable concept.

[9] Intermediate inputs and labour are linked to overall electricity production using fixed input coefficients; they are not differentiated by type of fuel.

[10] It must be emphasized that in the applications to be described the model is rather well-behaved with respect to heat rates, because the parameters have been calibrated to empirical data. This implies that the heat rates observed in the base year are reproduced by the base year model solution. For the baseline projections the efficiency coefficients α_i have been raised in such a way as to produce technically reasonable future heat rates.

$$(4)\quad FEL_i = EL_i + \left(\frac{EL_i}{a_i \left(K_i^{'}\right)^{1-b_i}} \right)^{1/b_i}$$

Thus, for any electricity output the required fuel input becomes smaller as the capital intensity and the capital efficiency increase. Because the capital intensity depends on the fuel price (see below), the (long-term) fuel input is a function of the fuel price as well.

The capital stock in existence at the beginning of each period is the result of a cost-minimising investment calculus carried out in the preceding period. Under static price expectations and in the absence of adjustment costs, the optimal investment program is obtained from the condition $-\partial VC_i/\partial K_i' = uc_i$, where VC_i refers to the variable cost function and uc_i' is the user cost of (effective) capital. In the case under consideration this can be solved to yield the planned capital stock as follows

$$(5)\quad KP_i^{'} = \frac{1}{\alpha_i} \left(\frac{1-b_i}{b_i} \right)^{b_i} \left(\frac{PFEL_i}{uc_i} \right)^{b_i} E(EL_i)$$

where $PFEL_i$ = price of fuel i, and $E(EL_i)$ = expected demand for electricity of type i (obtained from extrapolation).

For nuclear and renewable electricity the required input of capital and fuel is linked to output via fixed coefficients.[11] Other inputs to the electricity industry (labour and intermediates) are assumed to be proportional to *overall* electricity supply.

The demand for input factors is one major group of electricity variables, to be fed back into the general equilibrium model. Another important electricity variable is the price of electricity, taken to be the average cost. The average cost is obtained from cost functions dual to the quantity aggregator functions described above. Thus - given the input prices and the overall electricity demand, as provided by the general equilibrium model - the price of electricity depends on the composition of overall electricity in terms of fuel, and on the capital-to-fuel ratio for each type of electricity. Just as the demand for input factors, the electricity price is fed back into the general equilibrium model.

2.3 Data and Parameterisation

The model is calibrated against input-output tables (EUROSTAT 1992) and electricity data (IEA 1992) for 1985. The key exogenous drivers (i.e., world energy prices, nuclear capacity, renewable energy, domestic hard coal mining) 1985-1998 have been set according to their historical development. The elasticity assumptions used are given in Table 1. Of particular interest for the current application of the model are the electricity-related elasticity parameters, i.e. the elasticity of substitution between domestic and imported electricity, and the elasticity of substitution among the various fossil fuels. Our standard assumption for both

[11] The constant capital coefficient for nuclear refers to capital in 'effective units'. The efficiency of capital increases in time similar to that of capital in fossil-based generation.

of these parameters is 0.2. We assume that the low inter-fuel elasticity will continue to prevail in the future, reflecting a preference for a balanced fuel mix and a high degree of supply security. Considerations of supply security also justify assuming a low elasticity of domestic vs. imported electricity.

Table 1: Elasticity parameters

Armington elasticities:	
export of ROW vs. export of EC	0.1-0.8
export of GER vs. export of EC9	0.1-2.0
domestic products vs. imports	0.1-2.0
Substitution elasticities in production:	
energy/capital vs. labour	0.6
energy vs. capital	0.3-0.7
electricity vs. fuel	0.6
fuel vs. fuel	0.8
Electricity-related elasticities:	
domestic vs. imported electricity	0.2
fossil fuel vs. fossil fuel in electricity generation	0.2
Reaction elasticities:	
exchange rate w.r.t. current account ratio	0.2
savings ratio w.r.t. interest rate	0.8
wage rate w.r.t. employment	1.5

3. Assumptions and Scenarios

In this section we describe our assumptions, define the scenarios considered in our simulations, and provide a characterisation of the base case.

3.1 Assumptions

The key exogenous variables which determine the overall dynamics of the model are the rate of autonomous labour efficiency improvement and the rate of autonomous energy efficiency improvement. They are set at 1.8 percent and 1.5 percent, respectively. The exogenous energy-related variables refer to the evolution of world energy prices and to the use of domestic hard coal, renewable energy and nuclear electricity. The scenarios concerning nuclear energy will be considered below; the assumptions concerning the other exogenous energy-related variables are given in Table 2.

Table 2: Assumptions relating to energy (percentage annual change)

	2000-2010	2010-2020	2020-2030
Coal price	1.8	1.0	1.0
Oil price	2.4	1.8	2.3
Gas price	2.0	2.1	2.7
Domestic hard coal	-7.2	-5.5	-3.5
Renewable energy	3.7	2.1	1.8

Source: Forum für Energiemodelle und energiewirtschaftliche Systemanalysen (1999, Chapter 2)[12]

These assumptions entail a doubling of renewable energy and a dismantling of domestic hard coal by more than 80 percent over the next three decades. Any demand for hard coal will then largely be satisfied from imported coal.

3.2 Scenarios

We consider two scenarios of nuclear power generation:

Base Case: Existing nuclear plants are used over a lifetime of 40 years; there are no additions to nuclear capacity.

Phase-out Case: Existing capacities are used over a lifetime of 32 years; there are no capacity additions.

The evolution of nuclear capacity in these scenarios is depicted in Figure 1.

[12] The 'Forum' is an informal association of German energy economists, comparable in its objectives to the Energy Modeling Forum regularly organized by Stanford University. The projections developed by the 'Forum' are the basis of several joint studies of German energy policy issues, which is why they are used here.

Figure 1: Evolution of nuclear capacity

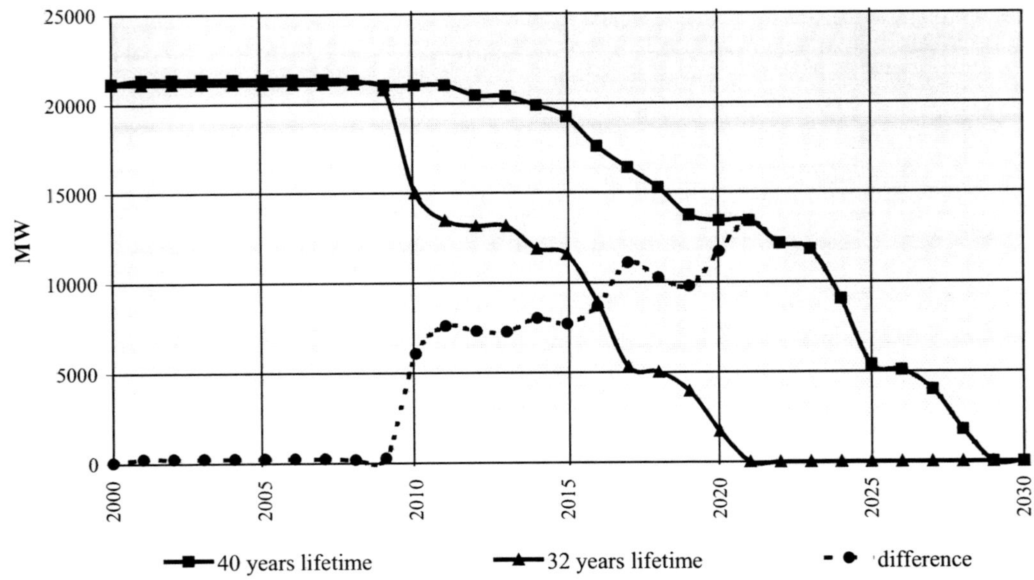

3.3 The Base Case

The base-case evolution of some key variables under the assumptions described above is presented in Table 3.

Table 3: Evolution of key variables in the base case (percentage annual change)

	2000-2010	2010-2020	2020-2030
GDP	2.0	1.9	1.8
Electricity generation	1.3	1.0	0.9
- Hard coal based	2.0	2.9	2.9
- Brown coal based	2.3	3.1	3.1
- Oil based	2.0	2.8	2.8
- Gas based	2.1	2.9	2.8
Electricity price	-0.2	0.3	0.5
CO_2	0.7	1.1	1.0

It can be seen that the base case entails a rather strong increase in fossil-fuelled electricity generation, as compared to the evolution of overall electricity generation. This effect arises, because the base-case capacity evolution implies that nuclear power is lagging behind the growth of demand. In the first decade, nuclear capacity is more or less constant while total electricity demand rises; in the second and third decade there is a decline in nuclear capacity while overall electricity demand continues to rise. The resulting supply gap is filled by additions to fossil-fuelled capacity. The enhanced requirement for capacity additions in the second and third decade triggers an increase in the electricity price, which explains why the electricity/GDP ratio drops faster in the second and third decade than in the first decade. With respect to CO_2 emissions, the increasing share of fossil-based electricity leads to higher growth of economy-wide CO_2 in the second and third decade. The evolution of the different fossil fuels is rather balanced, reflecting our assumption of a rather low elasticity of the fuel mix with respect to fuel prices.

4. EFFECTS OF NUCLEAR PHASE-OUT

This section describes the effects on the electricity industry and the overall economy of cutting down the lifetime of nuclear plants from 40 to 32 years.

A general feature of our simulations is that many of the direct, electricity-related effects of nuclear phase-out increase in magnitude until about 2020 and start decreasing rapidly. This reflects the increasing divergence of the two capacity trajectories until 2020 and the ensuing re-convergence, as displayed in Figure 1. This figure also shows that the removal of nuclear capacity will be particularly pronounced around 2010, which leads to stron economic effects around that year.

Another general result is that some of the macroeconomic and macro-sectoral effects change their sign over the time span, because they are the net result of counteracting partial effects whose magnitudes vary in time.

4.1 Electricity Industry

Table 4 shows the effect of the nuclear phase-out on electricity generation and electricity prices. All effects are expressed in terms of percentage changes from the base case. By 2010 the price of electricity is about 7 percent higher in the phase-out case than in the base case, because electricity generation costs are driven up by the capital costs related to the 'premature' replacement of nuclear capacity. The price differential drops somewhat by 2015, because there are few replacement requirements between 2010 and 2015 (see Figure 1). It increases to about 8 percent by 2020, and then declines drastically as the phase-out capacity trajectory approaches the base case. Actually, by 2025 the phase-out entails a slight fall in the electricity price. This result arises because by that time the amount and, hence, cost of 'premature' capacity replacement is small, whereas non-nuclear electricity generation becomes less expensive (due mainly to the moderate development of the coal price).

The price increase implies a fall in electricity demand and, hence, generation by up to 3.7 percent in 2020, which then declines strongly. The supply gap which remains after the reduced demand is accounted for is filled by increased fossil-based power generation, in which the various fuels play a more or less proportionate part. The latter reflects our assumption of a rather small elasticity of inter-fuel substitution.

Table 4: Effects of phase-out on electricity sector (percentage difference from base case)

	2000	2005	2010	2015	2020	2025	2030
Total generation	-0.02	-0.08	-2.90	-2.45	-3.71	-0.26	0.00
Hard coal based	0.10	0.51	9.74	11.68	15.14	7.30	0.04
Brown coal based	0.07	0.49	7.36	11.20	13.56	7.27	-0.05
Petroleum based	0.09	0.52	8.97	11.60	14.69	7.16	0.00
Gas based	0.10	0.51	10.12	11.65	15.33	7.27	-0.08
Electricity price	0.06	0.15	7.19	4.40	8.08	-0.29	0.15

4.2 Macroeconomic Effects

The macroeconomic effects of nuclear phase-out are indirect ones which basically result from two more direct, partial effects: the requirement for additions to fossil-fuelled capacity and the change in electricity prices. The relative importance of the two partial effects varies as the magnitude of capacity addition requirements changes over time.

The macroeconomic impacts, along with the effect on CO_2 emissions, are displayed in Table 5. Considering GDP, we find that the effect changes its sign over time. The effect is slightly positive around 2010 and 2020, i.e. the years when the expansionary impulse from capacity additions is large, leading to a boost for the equipment goods and construction sector (see section 4.3). In other periods, the electricity price effect dominates the capacity addition effect, leading to a decrease of GDP by up to 0.5 percent.

The effect on employment largely mirrors the GDP effect, but is more pronounced This is understandable, because the employment effect is more dependent on the required capacity additions than is GDP, and is therefore more volatile.

Regarding the environmental effect, it remains to mention that the nuclear phase-out implies an increase in economy-wide CO_2 emissions by up to 5.5 percent relative to the base case. This increase in emission rates also fades off as the two nuclear trajectories get closer to each other. However, due to the long atmospheric lifetime of CO_2, the transitory increase in emission rates will leave a permanent trace in terms of atmospheric concentration.

Table 5: Macroeconomic and environmental effects (percentage difference from base case)

	2000	2005	2010	2015	2020	2025	2030
GDP	0.00	-0.02	0.21	-0.52	0.17	-0.41	0.05
Employment	0.00	-0.03	0.58	-0.70	0.60	-1.15	0.26
CO_2-emissions	0.04	0.12	4.34	2.94	5.47	1.88	0.14

4.3 Sectoral Effects

The sectoral effects of the nuclear phase-out depend on the relative magnitude of the capacity addition requirements and the electricity price effect, as discussed above, and on the degree to which a particular industry is affected by these two kinds of impulse.

One way of assessing how susceptible an industry is to higher electricity prices is in terms of the overall index of energy prices facing this industry.[13] The induced changes in the energy price index of the non-energy industries are presented in Table 6. These changes display considerable dispersion across industries, the construction industry and transportation being affected much less than agriculture, the manufacturing sector, and the service sector. This reflects, of course, the relative electricity intensity of these industries.

Table 6: Effects on sectoral energy prices (percentage difference from base case)

	2000	2005	2010	2015	2020	2025	2030
Agriculture	0.02	0.06	2.70	1.60	2.88	-0.12	0.00
Intermediate goods	0.02	0.05	2.25	1.33	2.38	0.00	0.00
Equipment goods	0.04	0.09	3.95	2.37	4.29	-0.12	0.11
Consumption goods	0.03	0.08	3.63	2.17	3.92	-0.12	0.00
Construction	0.01	0.01	0.61	0.31	0.56	-0.11	0.00
Transportation	0.01	0.03	1.39	0.79	1.42	-0.11	0.00
Services	0.03	0.08	3.64	2.18	3.94	-0.12	0.11
Government	0.03	0.07	3.47	2.07	3.74	-0.12	0.11

When we consider the effects on outputs in the non-energy sectors (see Table 7), it is difficult to detect a straightforward pattern of influence from higher energy prices to changes in output; changes in output are mainly the result of inter-industry demand effects, which mostly follow a complicated pattern. There do, however, exist straightforward demand effects from required additions to fossil-fuelled electricity generation capacity to the output of the equipment goods industry and the construction sector. In those years in which capacity additions are large, these demand effects clearly dominate the cost burden that comes from higher electricity prices; therefore these two sectors are temporal gainers from the nuclear phase-out scenario. In terms of their magnitude, the sectoral output effects are rather moderate for most industries.

[13] A more comprehensive indicator of 'vulnerability' by higher electricity prices would be provided by the change of an industry's total unit cost. However, total unit costs are dominated by other influences, especially the wage rate and the prices of intermediates, which include all the general equilibrium effects taking place in response to nuclear phase-out. Therefore, we consider the change in the energy price index to be a more immediate and intuitive indicator.

Table 7: Effects on sectoral output (percentage difference from base case)

	2000	2005	2010	2015	2020	2025	2030
Agriculture	0.00	-0.01	-0.16	-0.29	0.00	-0.19	-0.15
Intermediate goods	0.00	-0.02	-0.16	-0.43	0.30	-0.08	-0.05
Equipment goods	0.01	-0.02	0.70	-0.59	0.77	-0.42	0.17
Consumption goods	0.00	-0.02	-0.04	-0.39	0.28	-0.17	-0.03
Construction	0.02	-0.03	2.60	-0.82	0.84	-1.37	0.63
Transportation	0.00	-0.02	0.17	-0.41	0.19	-0.28	0.03
Services	0.00	-0.02	0.27	-0.45	0.13	-0.47	0.01
Government	0.00	-0.01	-0.19	-0.33	0.08	-0.26	-0.01

5. Conclusions

This paper has studied the sectoral and macroeconomic effects of a nuclear phase-out in Germany. More specifically, the effects have been examined of cutting the lifetime of existing nuclear power plants from 40 years (base case) to 32 years (phase-out case), as recently agreed by the German government and electricity industry. Our method of analysis was to use a computable general equilibrium model which encompasses a detailed representation of the electricity supply industry.

A key feature of our results is that most effects of the nuclear phase-out (in comparison with the base case) are transitory because the trajectories of nuclear capacity which characterise the two scenarios approach each other towards the end of the time span considered. The maximum 'nuclear capacity gap' occurs in 2021.

The nuclear phase-out entails an increase of the electricity price by up to 8 percent and an ensuing fall in electricity demand by 3.7 percent. The supply gap that remains after the fall in demand is accounted for is filled by additions to fossil-fuelled capacity. Assuming a low degree of inter-fuel substitution, these additions affect hard coal, brown coal and natural gas more or less proportionately.

The macroeconomic effects of the phase-out display rather pronounced dynamics: In those years in which the need for additional fossil-fuelled generation capacity is large, there is a small expansionary effect for GDP and employment. When replacement requirements are smaller the electricity price effect plays the dominating role, and the effect on GDP and employment turns negative. As the phase-out nuclear trajectory approaches the base-case trajectory, the macroeconomic effects fade off. In terms of magnitude, the effect on GDP is always less than 0.5 percent, and much less than that most of the time. The economy-wide CO_2 emissions rise by up to 5.5 percent in 2020, but decline again to the base case level until 2030.

Attempting an overall assessment of the nuclear phase-out initiated in Germany, it can be said that the effects outside the electricity industry are mostly rather moderate. A notable and important exception to this general statement is the significant enhancement of CO_2 emissions; this will certainly make it harder for Germany to meet Kyoto (and prospective post-Kyoto) obligations.

REFERENCES

Armington, P.S. (1969), A Theory of Demand for Products Distinguished by Place of Production, *IMF Staff Papers* 16, 159-178.

Burniaux, J.-M., Martin, J.P., Nicoletti, G., Olivera Martins, J. (1992), *A Multi-Sector, Multi-Region General Equilibrium Model for Quantifying the Costs of Curbing CO_2 Emissions: A Technical Manual*, OECD Economics Department Working Paper No. 116, Paris.

EUROSTAT (1992), *National Accounts: Input Output Tables* 1985, Luxembourg.

Forum für Energiemodelle und Energiewirtschftliche Systemanalysen (1999), *Energiemodelle zum Klimaschutz in Deutschland: Strukturelle und gesamtwirtschaftliche Auswirkungen aus nationaler Perspektive*, Heidelberg.

Hoster, F. (1998), Impact of a Nuclear Phase-out in Germany: Results from a Simulation Model of the European Power Systems, *Energy Policy* 26, 507-518.

Hoster, F., Welsch, H., Böhringer, C. (1997), *CO_2 Abatement and Economic Structural Change in the European Internal Market*, Heidelberg.

IEA (1992), *Electricity Information*, Paris.

Pfaffenberger, W. (1999), *Ausstieg aus der Kernenergie - und was kommt danach?*, Frankfurt.

Pfaffenberger, W., Gerdey, H.-J. (1998), *Zur Bedeutung der Kernenergie für die Volkswirschaft und die Umwelt. Zur Abschätzung der Kosten eines Ausstiegs*, Bremer Energie-Institut, Bremen.

Welsch, H. (1998), Coal Subsidization and Nuclear Phase-out in a General Equilibrium Model for Germany, *Energy Economics* 20, 203-222.

Welsch, H., Hoster, F (1995), A General Equilibrium Analysis of European Carbon/Energy Taxation: Model Structure and Macroeconomic Results, *Zeitschrift für Wirtschafts- und Sozialwissenschaften 115*, 275-303.

Chapter 5

CONTROL OF SPECIFIC ENERGY CONSUMPTION AND ENERGY EFFICIENCY IN SELECTED INDUSTRIES FOR CO_2 MITIGATION

M. Siddhartha Bhatt and Vinita Chellappan
Central Power Research Institute, Energy Research Centre,
Trivandrum-695017, India

ABSTRACT

This paper presents a novel demand side management (DSM) approach for improving energy efficiency and mitigating CO_2 emissions- maximisation of the instantaneous production level in an industry while maintaining a required overall production level. The specific energy consumption (SEC) in an industry varies inversely as the instantaneous production rate, the factor of decrease in the SEC being as high 1/3 to 1/4. To take advantage of this fact, the no load (zero production level) energy consumption of the plant must be minimised to around 5-10 % from the present 25-50 % by measures such as matching and sizing of equipment capacity, operational optimisation, installing efficient controls in the utilities, etc.. It is concluded that a re-thinking on the plant operation vis-à-vis the energy consumption is required. Significant reduction in energy consumption and CO_2 emissions is technically feasible if the plant is operated in an optimal production on-production off sequence (thereby achieving a given production level as dictated by the market) while maintaining the maximum production rate during the production on period.

Key words: Specific energy consumption, production level sequencing.

INTRODUCTION

Mitigation of CO_2 through improved energy efficiency can be tackled on the supply side as well as on the demand side. Sivaramankrishnan and Siddiqi (1997) have estimated CO_2 generation for the Indian power sector and have suggested three strategies to limit these emissions: energy efficient generating technologies, shift to low or nil carbon energy sources and improvement in end use efficiency. Painuly et al. (1997) have focussed on CO_2 reduction by minimising auxiliary power in coal fired thermal plants as a supply side strategy. Gellings (1997) has identified motor drives, electrolysis, process heating and lighting as the areas with potential for enhancing energy efficiency and mitigating CO_2 emissions. His approach is essentially a demand side management (DSM) one- the widespread change-over to electric technologies such as adjustable speed drives, plasma fired cupolas, freeze concentration evaporation, ultrasound drying , etc., for improved energy efficiency.

In this paper, a novel and simple DSM methodology for improvement in energy efficiency through maximisation/optimisation of the instantaneous production rate in an industry is proposed. The main energy consumption in an industrial plant can be accounted for by two distinct sub-systems:

- **Utilities**: These provide a centralised source of energy or energised fluid to the whole plant. Utilities consist of compressed air plants, chilled water plants, steam generators, compressed water plants, diesel generating sets, electrical distribution networks, etc..
- **End use equipment**: These directly consume electrical energy and fuel. They consist of thermal equipment such as heaters, kilns, ovens, etc., and electro-mechanical machinery such as machine tools, production machinery, etc..

The specific energy consumption (SEC) which is the energy consumption to produce a unit of product (expressed in kWh/kg of product, kg of solid fuel/kg of product or litre of fuel oil/kg of product) varies inversely with increase in production rate. It is a minimum at the maximum production rate. The plant cannot be always operated at the point of the minimum SEC because considerations of product off take in the market, future level of orders and existing inventory levels dictate the production rate. Also, the design full load capacity of the plant may not coincide with the point of minimum SEC for the following reasons:

- The end use equipment in the plant which account for major energy consumption would have been installed independent of each other and at different points of time. The design margins in their capacity could be different.
- The sizing of utilities would have varying design margins and redundancies. During plant operation at 100 % plant capacity, their individual capacity factors would be low.
- Equipment addition/deletion/upgradation/downsizing is a continuous process and this is often done on considerations of importance to a local sub-system without considering the overall system.
- Production sub-systems are not fully synchronised because of equipment limitations.

The SEC is high at low production levels because:

- The constant energy consumption components of the plant (lighting, heating, ventilation, etc.,) are high in proportion to the variable components.
- Over operation of utilities result in wastage or re-circulation flows.
- Part load efficiency of energy intensive equipment such as mills, grinders, etc., are low.
- Inadequate controls/communication between energy sources and the load result in wastage.
- Insensitivity of batch type equipment to loading such as kilns, heaters, etc., produce no reduction in energy consumption at part loads.
- In case the production level is hindered by frequent breakdowns and outages, energy wastage is imminent.

Specific CO_2 generation follows the same trend as that of the SEC. If the SEC is low and the plant is operated at the optimal production level, then the specific CO_2 will be minimum.

ANALYSIS

The indices of energy consumption and CO_2 production are described below.

i. If the plant SEC data is plotted as a function of the capacity factor (CF) (range 0 to 1), a scatter of points is obtained. A linear best fit curve can then be fitted to the data in the form,

$$SEC = a_0 + a_1 CF \qquad (1)$$

where a_0 and a_1 are constants, and CF is the capacity factor.

In this paper, the SEC of industrial products are expressed in terms of either unit weight of fuel or unit electrical energy consumption. The conventional unit of SEC is energy based (e.g., kJ/kg, etc.,) (Boustead and Hancock, 1979; Perry and Green, 1984). This unit combines energy sources with different quality levels such as electrical energy and fuel energy which renders it unfeasible to precisely determine the CO_2 generation indices. Hence, in this study, the contributions due to electrical and fuel energy are de-coupled and treated separately.

ii. Using these constants, the SEC ratio (SECR) is defined as,

$$SECR = [\, a_0/(a_0+a_1)] \quad (2)$$

The SECR is the ratio of the maximum SEC to the minimum SEC under the assumption of a linear relationship between the SEC and the plant capacity.

iii. If the normalised energy consumption (with reference to the energy consumption at full capacity) (NEC) is plotted as a function of the capacity factor (0-1) and a best fit line is obtained, it will be of the form,
NEC = $b_0 + b_1$(CF) (3)

where the constants b_0 and b_1 represent the no load energy component and the component which is sensitive to the plant capacity, respectively. CF is the capacity factor. The following is to be noted:

- $b_0 + b_1 = 1.0$.
- b_0 consists of three components:
  - Constant component of energy of the industry which will be incurred whether the production processes are on or off.
  - The no load energy consumption of end use equipment.
  - The no load energy consumption of utilities in active service (energised and in hot stand-by) but at zero production level.

iv The CO_2 generation index ($I_{F\text{-}CO2}$) of a fuel is given by,

$$I_{F\text{-}CO2} = f_C(44/12) \quad (4)$$

where f_c is the fraction of carbon in fuel.

v. The CO_2 generation index of electrical energy delivered at the industrial site is given by,

$$I_{E\text{-}CO2} = [\, f_C(44/12)\, 3600] / [(f_C C_C + f_{H2} C_{H2} + f_S C_S)\eta_E \eta_A \eta_T] \quad (5)$$

where f_C, f_{H2}, f_S are the fraction of carbon, hydrogen and sulphur in the fuel. C_C, C_{H2} and C_S are the calorific values of carbon, hydrogen and sulphur. η_E, η_A and η_T are the efficiencies of gross electric conversion, auxiliary power in the power plant and transmission cum distribution. A comparison of the range of efficiencies in the developed countries and in Indian conditions is given in Table 1.

Table 1: Comparison of overall electric generation, transmission and distribution efficiencies.

Sl. No.	Particular	Developed countries	Indian conditions
01	Gross overall efficiency	38-45 %	24-32 %
02	Auaxiliary power efficiency	93-94 %	85-91 %
03	Transmission & distribution efficiency	94-95 %	75-80 %
04	Overall electric efficiency	33-40 %	16-23 %

vi The specific CO_2 generation index ($SI_{E\text{-}CO2}$) of an industrial product consuming electrical energy is given by,

$$SI_{E\text{-}CO2} = SEC\ I_{E\text{-}CO2} \quad (6)$$

vii The specific CO_2 generation index ($SI_{F\text{-}CO2}$) of an industrial product consuming fuel energy is given by,

$$SI_{F\text{-}CO2} = SEC\ I_{F\text{-}CO2} \quad (7)$$

viii The increased specific CO_2 generation ($SII_{E\text{-}CO2}$) for an industrial product (consuming electrical energy) due to part load operation of the plant is given by,

$$SII_{E\text{-}CO2} = SECR\ I_{E\text{-}CO2} \quad (8)$$

ix The increased specific CO_2 generation ($SII_{F\text{-}CO2}$) for an industrial product (consuming fuel energy) due to part load operation of the plant is given by,

$$SII_{F\text{-}CO2} = SECR\ I_{F\text{-}CO2} \quad (9)$$

EXPERIMENTAL WORK

Energy audits have been carried out in a large number of industries and measurement of energy inputs, production rates, etc., have been undertaken. Power measurement is by 3-phase power profilers (accuracy: 0.2 %). Fuel is measured by level indicators (accuracy: 5.0 %).

Errors in data can occur possibly due to some of the following causes:

- Changes in the accounting period and carry over of either fuel, energy or products to the next accounting interval for a variety of reasons.
- When there is a heavy demand for products or components, the industries sometimes off-load work to outside agencies and get the products from external sources. In such a case, the energy consumption of the off-loaded products are not accounted in the industry's balance sheet.
- In the event of an entire lot being rejected by the bulk buyers, these lots are re-cycled with 100 % quality checks. The products passing the quality checks are re-sold.
- Faulty metering.

Though every effort is made to minimise errors, fool proof data acquisition is quite difficult in industrial systems. However, regression data is provided to minimise the effect of errors in the SEC.

While attempt is made to have representative samples of all industries, the choice of the industries is purely based on the willingness of the industries to conduct studies in their premises. Hence, there are more samples of certain industries like cement and less of others like sugar. All industries studied over the past 2-3 years are reported in this paper. The number of industries reported here have no bearing on their national market share, energy share, etc..

Results and Discussion

Table 2 gives the CO_2 generation indices for Indian fuels and electrical energy generated from the coal fired thermal stations. Otfried Voigt (1995) has estimated the CO_2 emission from fossil fuel plants as follows:

- Fuel oil: 0.76 kg/kWh
- Hard coal-steam plant: 0.82 kg/kWh
- Hard coal- combined cycle plant with integrated coal gasification: 0.79 kg/kWh
- Lignite-steam plant: 1.04 kg/kWh

Table 2: Specific CO_2 generation index for fuel combustion and electric production.

Sl. No.	Particular	Fuel combustion	Electric production
		Specific mass -CO_2 production (kg/kg)	Specific electricity - CO_2 production (kg/kWh)
01	Fuel oil: light diesel oil/ fuel oil	3.15	0.95
02	Fuel oil: high speed diesel/ heavy fuel oil	3.10	0.92
03	Fuel oil: low sulpher heavy stock	3.03	0.97
04	Bituminus coal	1.46	1.13

The Indian values of CO_2 generation indices are higher because of the lower efficiencies of generation, transmission and distribution; and higher auxiliary power.

It is seen from the formulae in the section on analysis, that the specific CO_2 generation indices of the industrial products are dependant on the SEC of the process in the industry as well as on the overall conversion chain efficiency of the utility. The approach here is to control the SEC of the products. A novel parameter is adopted to quantify the extent of CO_2 mitigation due to optimisation of the production levels: SEC ratio.

Figures 1-8 shows plots of variation of SEC with CF for a few typical cases. It can clearly be seen that as the production level increases, the SEC decreases. The energy efficiency is best when the plant is operated at the maximum production level.

Tables 3-9 gives the constants a_0 and a_1 for SEC of various industries based on experimental measurements over periods ranging from 6 months to 18 months. The maximum plant capacities are also shown in the Tables. Wherever the constant a_1 is 0, it implies that the variation of SEC with production rate could not be obtained because of the following causes:

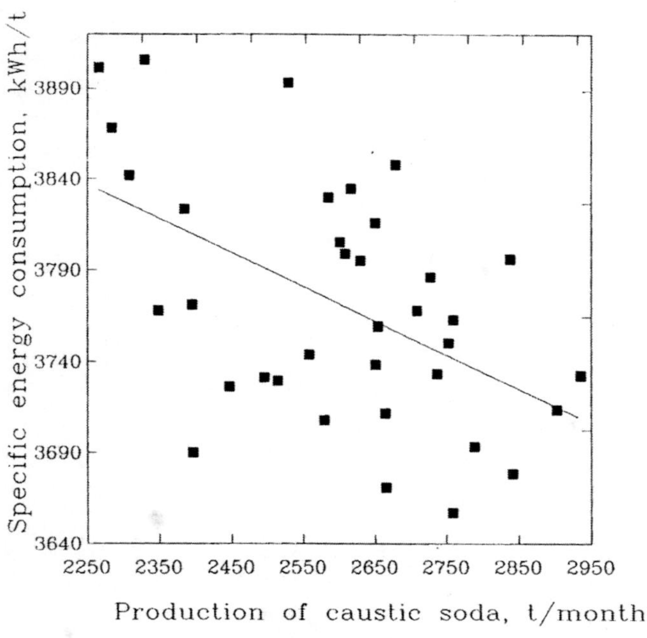

Figure 1: Variation of SEC of caustic soda with production.

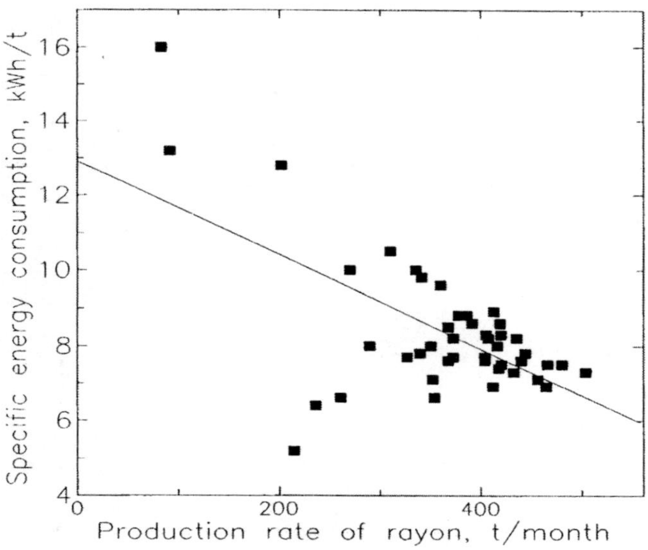

Figure 2: Variation of SEC of rayon with production.

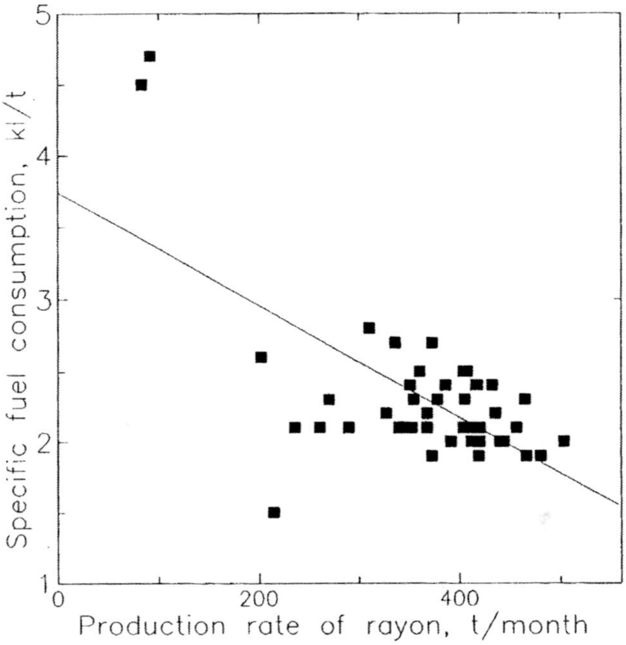

Figure 3: Variation of SFC of rayon with production.

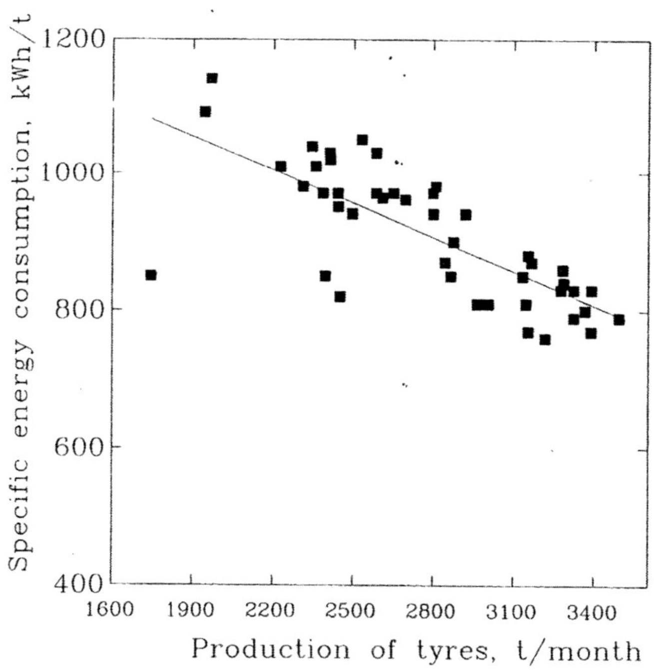

Figure 4: Variation of SEC of tyres with production.

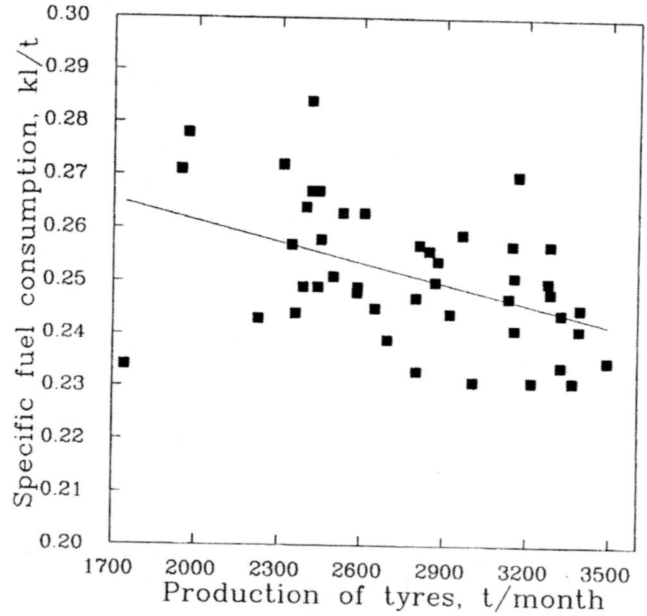

Figure 5: Variation of SFC of tyres with production.

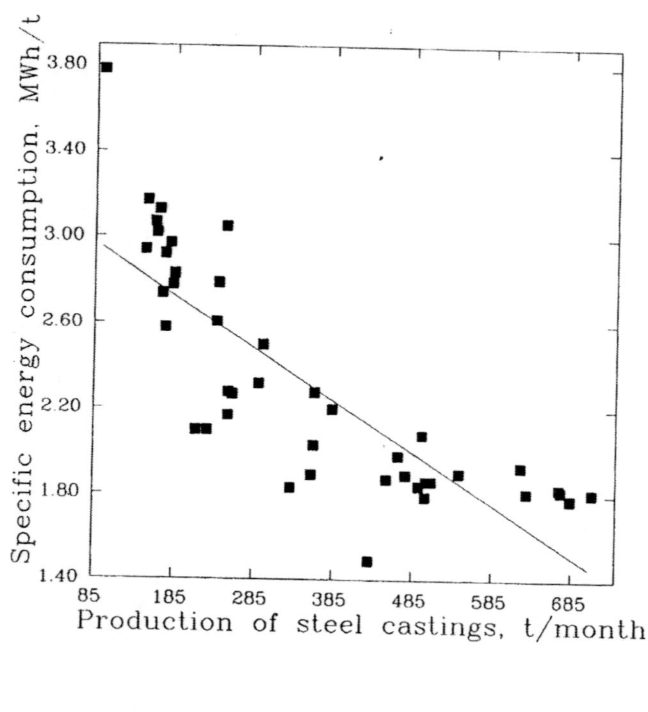

Figure 6: Variation of SEC of steel castings with production.

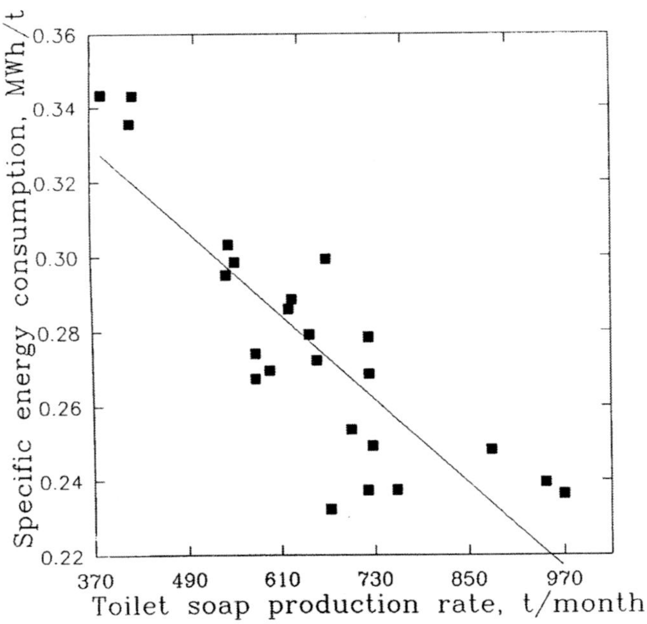

Figure 7: Variation of SEC of toilet soap with production.

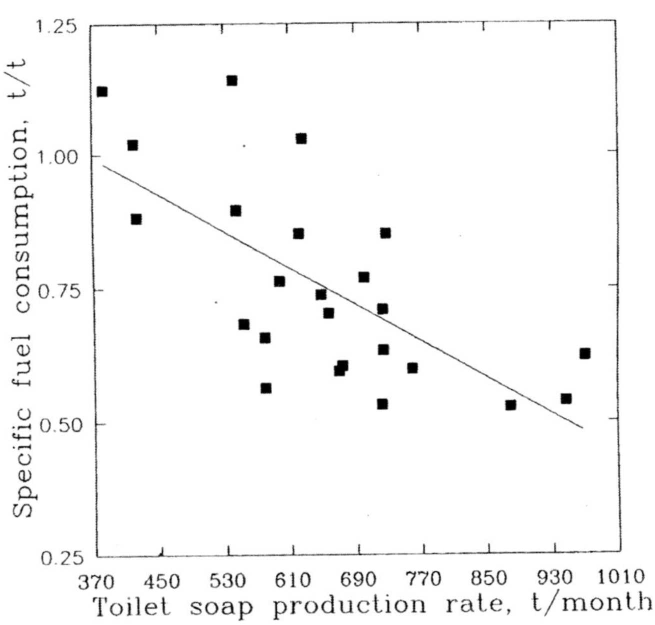

Figure 8: Variation of SFC of toilet soap with production.

Table 3: Constants in the curve fits for specific energy consumption (SEC) for chemical products

Sl. No.	Chemicals Range of X= 0 to 1	a_0	a_1
01	X: Styrene	Max. X = 1000 t/month	
	Y: Specific electrical consumption of plant (kWh/t)	1387.000	-886.710
	Y: Specific fuel oil consumption of plant (l/t)	334.450	0.0
02	X: Polystyrene	Max. X = 1000 t/month	
	Y: Specific electrical consumption of plant (kWh/t)	244.010	-44.490
	Y: Specific fuel oil consumption of plant (l/t)	23.160	0.0
03	X: PVC resins and compounds	Max. X = 450 t/month	
	Y: Specific electrical consumption of plant (kWh/t)	5407.650	-254.720
	Y: Specific fuel oil consumption of plant (l/t)	20.050	-4.309
04	X: Linear alkyl benzene	Max. X = 7000 t/month	
	Y: Specific electrical consumption of plant (kWh/t)	364.330	-58.460
	Y: Specific fuel oil consumption of plant (l/t)	64.500	0.0
05	X: High/low density polyethylene	Max. X = 1300 t/month	
	Y: Specific electrical consumption of plant (kWh/t)	434.700	-238.030
	Y: Specific coal consumption of plant (kg/t)	47.700	-15.370
06	X: Caustic soda	Max. X= 3000 t/month	
	Y: Specific electrical consumption of plant (kWh/t)	4257.280	-560.451
07	X: Sodium silicate	Max. X = 250 t/month	
	Y: Specific fuel oil consumption of plant (l/t)	165.667	-26.667

Table 4: Constants in the curve fits for specific energy consumption (SEC) for cement

Sl. No.	Cement Range of X= 0 to 1	a_0	a_1
01	X: White cement	Max. X = $0.5x10^6$ t/year	
	Y: Specific electrical consumption of plant (kWh/t)	183.260	-62.040
	Y: Specific fuel oil consumption of plant (l/t)	130.206	0.0
02	X: Portland cement plant # 01	Max. X = $0.5x10^6$ t/year	
	Y: Specific electrical consumption of plant (kWh/t)	117.860	-12.980
	Y: Specific fuel oil consumption of plant (l/t)	120.055	-40.000
03	X: Portland cement plant # 02	Max. X = $0.5x10^6$ t/year	
	Y: Specific electrical consumption of plant (kWh/t)	99.998	-10.290
04	X: Portland cement plant # 03	Max. X = $0.5x10^6$ t/year	
	Y: Specific electrical consumption of plant (kWh/t)	130.010	-47.940
05	X: Portland cement plant # 04	Max. X = $0.5x10^6$ t/year	
	Y: Specific electrical consumption of plant (kWh/t)	141.540	-35.240
06	X: Portland cement plant # 05	Max. X = $0.25x10^6$ t/year	
	Y: Specific electrical consumption of plant (kWh/t)	136.060	-37.570
07	X: Portland cement plant # 06	Max. X = $0.3x10^6$ t/year	
	Y: Specific electrical consumption of plant (kWh/t)	151.290	-57.670

Table 5: Constants in the curve fits for specific energy consumption (SEC) for paper and paper products

Sl. No.	Paper and paper products Range of X= 0 to 1	a_0	a_1
01	X: Paper plant # 01	Max. X = 3300 t/month	
	Y: Specific electrical consumption of plant (kWh/t)	1417.880	-111.320
02	X: Paper plant # 02	Max. X = 1500 t/month	
	Y: Specific electrical consumption of plant (kWh/t)	1427.500	-168.220
03	X: Printing paper	Max. X = 1500 t/month	
	Y: Specific electrical consumption of plant (kWh/t)	3003.160	-1492.800
	Y: Specific fuel oil consumption of plant (l/t)	100.300	-71.000
04	X: Cast coated paper	Max. X = 13400 t/month	
	Y: Specific electrical consumption of plant (kWh/t)	2004.280	-1231.200
	Y: Specific fuel oil consumption of plant (l/t)	193.000	-82.000
05	X: Paper board	Max. X = 9000 t/month	
	Y: Specific electrical consumption of plant (kWh/t)	2431.200	-1280.900
	Y: Specific fuel oil consumption of plant (l/t)	4.000	0.0

Table 6: Constants in the curve fits for specific energy consumption (SEC) for steel

Sl. No.	Steel Range of X= 0 to 1	a_0	a_1
01	X: Sponge iron	Max. X = 12500 t/month	
	Y: Specific electrical consumption of plant (kWh/t)	102.040	-7.830
02	X: Cold rolled stripes	Max. X = 2000 t/month	
	Y: Specific electrical consumption of plant (kWh/t)	403.860	-265.130
	Y: Specific fuel oil consumption of plant (l/t)	26.200	0.0
03	X: Carbon alloy steel	Max. X = 8300 t/month	
	Y: Specific electrical consumption of plant (kWh/t)	1227.210	-254.040
	Y: Specific fuel oil consumption of plant (l/t)	57.000	0.0
04	X: Steel	Max. X = 2000 t/month	
	Y: Specific electrical consumption of plant (kWh/t)	494.360	-25.410
05	X: Steel coils	Max. X = 3000 t/month	
	Y: Specific electrical consumption of plant (kWh/t)	1430.460	-407.510
	Y: Specific fuel oil consumption of plant (l/t)	56.000	0.0
06	X: Continuous automatic castings	Max. X= 1500 t/month	
	Y: Specific electrical consumption of plant (kWh/kg)	3.150	-1.610
	Y: Specific fuel oil consumption of plant (l/t)	27.528	-19.958

Table 7: Constants in the curve fits for specific energy consumption (SEC) for yarn and fabrics

Sl. No.	Yarn and fabrics Range of X= 0 to 1	a_0	a_1
01	X: Yarn plant # 01	Max. X = 1000 t/month	
	Y: Specific electrical consumption of plant (kWh/t)	3538.940	-1230.230
02	X: Yarn plant # 02	Max. X = 700 t/month	
	Y: Specific electrical consumption of plant (kWh/t)	4390.000	-78.500
03	X: Rayon fibre	Max. X= 550 t/month	
	Y: Specific electrical consumption of plant (kWh/t)	3.730	-1.917
	Y: Specific fuel oil consumption of plant (kl/t)	13.375	-7.560
04	X: Yarn plant # 03	Max. X = 500 t/month	
	Y: Specific electrical consumption of plant (kWh/t)	3890.000	-236.840
	Y: Specific coal consumption of plant (t/t)	7.308	-4.323
05	X: Synthetic fibre and yarn	Max. X = 500 t/month	
	Y: Specific electrical consumption of plant (kWh/t)	5171.880	-1948.000
06	X: Viscose filled yarn	Max. X = 1000 t/month	
	Y: Specific electrical consumption of plant (kWh/t)	6186.660	-60.460
	Y: Specific fuel oil consumption of plant (l/t)	124.400	-81.900
07	X: Cotton yarn	Max. X = 750 t/month	
	Y: Specific electrical consumption of plant (kWh/t)	3836.900	-325.890
	Y: Specific coal consumption of plant (t/t)	4.390	0.000
08	X: Fabrics plant # 01	Max. X = 400 km/month	
	Y: Specific electrical consumption of plant (kWh/km)	1450.300	-255.020
	Y: Specific fuel oil consumption of plant (l/km)	4.500	0.0
	Y: Specific coal consumption of plant (t/t)	0.859	0.0
09	X: Fabrics plant # 02	Max. X = 40 km/month	
	Y: Specific electrical consumption of plant (kWh/km)	923.900	-51.300
	Y: Specific coal consumption of plant (t/km)	1100.000	-383.000
10	X: Cloth	Max. X = 60 km/month	
	Y: Specific electrical consumption of plant (kWh/km)	2360.000	-1150.000
	Y: Specific coal consumption of plant (t/km)	0.650	0.0
11	X: Jute goods	Max. X = 3000 t/month	
	Y: Specific electrical consumption of plant (kWh/t)	941.780	-491.520
	Y: Specific coal consumption of plant (t/t)	379.000	0.0

Table 8: Constants in the curve fits for specific energy consumption (SEC) for food products

Sl. No.	Food products Range of X= 0 to 1		a_0	a_1
01	X: Bread (1 PU= 0.40 kg)	Max. X= 1.25E+06 units/month		
	Y: Specific electrical consumption of plant (kWh/1E+03 PU)		68.515	-36.311
	Y: Specific fuel oil consumption of plant (l/1.0E+03 PU)		31.133	-8.620
02	X: Sugar	Max. X = 7000 t/month		
	Y: Specific electrical consumption of plant (kWh/t)		526.030	-237.130
	Y: Specific fuel oil consumption of plant (l/t)		2.770	0.0
03	X: Edible oil plant # 01	Max. X = 2000 t/month		
	Y: Specific electrical consumption of plant (kWh/t)		159.800	-5.400
04	X: Edible oil plant # 02	Max. X = 1600 t/month		
	Y: Specific electrical consumption of plant (kWh/t)		89.810	-8.920
	Y: Specific coal consumption of plant (t/t)		0.163	0.0
05	X: Milk food	Max. X = 7000 t/month		
	Y: Specific electrical consumption of plant (kWh/t)		0.771	-0.210
	Y: Specific fuel oil consumption of plant (kg/t)		0.650	0.0
06	X: Milk products	Max. X = 800 t/month		
	Y: Specific electrical consumption of plant (kWh/t)		1.111	-0.624
	Y: Specific fuel oil consumption of plant (l/t)		1.150	0.0
07	X: Dairy products	Max. X = 1500 t/month		
	Y: Specific electrical consumption of plant (kWh/kl)		340.070	-180.600
08	X: Confectionery and food drinks	Max. X = 600 t/month		
	Y: Specific electrical consumption of plant (kWh/t)		1386.700	-244.540
	Y: Specific fuel oil consumption of plant (l/t)		346.200	-106.500
09	X: Ice cream	Max. X = 50 t/month		
	Y: Specific electrical consumption of plant (kWh/t)		2156.700	-121.000
10	X: Fruit pulp	Max. X = 200 t/month		
	Y: Specific electrical consumption of plant (kWh/t)		697.800	-391.600
	Y: Specific fuel oil consumption of plant (l/t)		71.500	-46.500

Note: PU: production unit.

Table 9: Constants in the curve fits for specific energy consumption (SEC) for miscellaneous products

Sl. No.	Miscellaneous Range of X= 0 to 1	a_0	a_1
01	X: Automobile tyre	Max. X= 3500 t/month	
	Y: Specific electrical consumption of plant (kWh/t)	1398.280	-637.880
	Y: Specific fuel oil consumption of plant (kl/t)	0.289	-0.0478
02	X: Kaolin	Max. X= 800 t/month	
	Y: Specific electrical consumption of plant (kWh/kg)	0.494	-0.226
	Y: Specific fuel oil consumption of plant (l/t)	151.439	-0.081
03	X: Bulk drugs	Max. X = 250 t/month	
	Y: Specific electrical consumption of plant (kWh/t)	136.370	-7.300
	Y: Specific fuel oil consumption of plant (l/t)	5.220	0.0
04	X: Toilet soap	Max. X= 2000 t/month	
	Y: Specific electrical consumption of plant (kWh/kg)	0.399	-0.188
	Y: Specific coal consumption of plant (t/t)	1.293	-0.793
05	X: Detergent	Max. X= 800 t/month	
	Y: Specific electrical consumption of plant (kWh/kg)	0.187	-0.138
06	X: Transparent poly vinyl blood bags	Max. X= 0.5E+06 PU/month	
	Y: Specific electrical consumption of plant (kWh/PU)	0.526	-0.342
	Y: Specific fuel oil consumption of plant (ml/PU)	62.530	-37.760
07	X: Condoms	Max. X= 15.0E+06 units/month	
	Y: Specific electrical consumption of plant (MWh/ 1.0E+06 PU)	17.131	-12.007
	Y: Specific fuel oil consumption of plant (kl/1.0E+06 PU)	12.025	-8.099
08	X: Ceramic crockery	Max. X= 50 t/month	
	Y: Specific electrical consumption of plant (kWh/kg)	17.789	-0.155
09	X: Super refractory	Max. X = 5000 t/month	
	Y: Specific electrical consumption of plant (kWh/t)	342.520	-159.680
	Y: Specific fuel oil consumption of plant (l/t)	518.000	0.0
10	X: Insulator	Max. X = 6000 t/month	
	Y: Specific electrical consumption of plant (kWh/t)	1019.480	-189.040
	Y: Specific fuel oil consumption of plant (kl/t)	2.260	-1.990

Note: PU: production unit.

- The variation in the operating production rate (experimentally measured) was so small that the SEC data could not be extrapolated over the entire operating range.
- The regression analysis did not show any correlation.
- There was physically no significant variation in the SEC with production rate.

It can be seen from the above Tables that, in general, the SEC is quite strongly sensitive to production rate. At a high production rate, it is possible to bring down the SEC.

Tables 10-16 gives constants b_0 and b_1 for NEC for the industries based on experimental data. The Tables show that the average no load electrical consumption (b_0) varies between 21 % and 50 % of the total energy consumption. The average no load fuel consumption (b_0) varies between 24 % and 57 % of the total consumption. This implies that when the plant is not producing, as much as 21-57 % of the energy is being consumed to maintain the plant as a hot stand-by. Ideally, the no load energy consumption must not exceed 5-10 % of the total plant consumption to account for lighting, ventilation and air conditioning. The values of b_0 are high for almost all major industrial sectors studied here.

Table 10: Constants in the curve fits for normalised energy consumption (NEC) for chemical products

Sl. No.	Chemicals Range of X= 0 to 1; Range of Y: 0 to 1	b_0	b_1
01	X: Styrene		
	Y: Normalised electrical consumption of plant	0.412	0.588
02	X: Polystyrene		
	Y: Normalised electrical consumption of plant	0.120	0.880
03	X: PVC resins and compounds		
	Y: Normalised electrical consumption of plant	0.334	0.666
	Y: Normalised fuel oil consumption of plant	0.154	0.846
04	X: Linear alkyl benzene		
	Y: Normalised electrical consumption of plant	0.164	0.836
05	X: High/low density polyethylene		
	Y: Normalised electrical consumption of plant	0.393	0.607
	Y: Normalised coal consumption of plant	0.648	0.352
06	X: Caustic soda		
	Y: Normalised electrical consumption of plant	0.157	0.843
07	X: Sodium silicate		
	Y: Normalised fuel oil consumption of plant	0.086	0.914
	Average value (electric consumption)	**0.263**	**0.737**
	Average value (fuel consumption)	**0.296**	**0.704**

Table 11: Constants in the curve fits for normalised energy consumption (NEC) for cement

Sl. No.	Cement Range of X= 0 to 1; Range of Y: 0 to 1	b_0	b_1
01	X: White cement		
	Y: Normalised electrical consumption of plant	0.374	0.626
02	X: Portland cement plant # 01		
	Y: Normalised electrical consumption of plant	0.115	0.885
	Y: Normalised fuel oil consumption of plant	0.460	0.540
03	X: Portland cement plant # 02		
	Y: Normalised electrical consumption of plant	0.096	0.904
04	X: Portland cement plant # 03		
	Y: Normalised electrical consumption of plant	0.488	0.512
05	X: Portland cement plant # 04		
	Y: Normalised electrical consumption of plant	0.306	0.694
06	X: Portland cement plant # 05		
	Y: Normalised electrical consumption of plant	0.315	0.685
07	X: Portland cement plant # 06		
	Y: Normalised electrical consumption of plant	0.548	0.452
	Average value (electric consumption)	**0.320**	**0.680**
	Average value (fuel consumption)	**0.460**	**0.540**

Table 12: Constants in the curve fits for normalised energy consumption (NEC) for paper and paper products

Sl. No.	Paper and paper products Range of X= 0 to 1; Range of Y: 0 to 1	b_0	b_1
01	X: Paper plant # 01		
	Y: Normalised electrical consumption of plant	0.167	0.833
02	X: Paper plant # 02		
	Y: Normalised electrical consumption of plant	0.138	0.862
03	X: Printing paper		
	Y: Normalised electrical consumption of plant	0.893	0.107
	Y: Normalised fuel oil consumption of plant	0.454	0.546
04	X: Cast coated paper		
	Y: Normalised electrical consumption of plant	0.651	0.349
	Y: Normalised fuel oil consumption of plant	0.425	0.575
05	X: Paper board		
	Y: Normalised electrical consumption of plant	0.684	0.316
	Average value (electrical consumption)	**0.507**	**0.493**
	Average value (fuel consumption)	**0.439**	**0.561**

Table 13: Constants in the curve fits for normalised energy consumption (NEC) for steel

Sl. No.	Steel Range of X= 0 to 1; Range of Y: 0 to 1	b_0	b_1
01	X: Sponge iron		
	Y: Normalised electrical consumption of plant	0.071	0.929
02	X: Cold rolled strips		
	Y: Normalised electrical consumption of plant	0.437	0.563
03	X: Carbon alloy steel		
	Y: Normalised electrical consumption of plant	0.218	0.782
04	X: Steel		
	Y: Normalised electrical consumption of plant	0.149	0.851
05	X: Steel coils		
	Y: Normalised electrical consumption of plant	0.203	0.797
06	X: Continuous automatic castings		
	Y: Normalised electrical consumption of plant	0.199	0.801
	Y: Normalised fuel oil consumption of plant	0.241	0.759
	Average value (electrical consumption)	**0.213**	**0.787**
	Average value (fuel consumption)	**0.241**	**0.759**

Table 14: Constants in the curve fits for normalised energy consumption (NEC) for yarn and fabrics

Sl. No.	Yarn and fabrics Range of X= 0 to 1; Range of Y: 0 to 1	b_0	b_1
01	X: Yarn plant # 01		
	Y: Normalised electrical consumption of plant	0.385	0.615
02	X: Yarn plant # 02		
	Y: Normalised electrical consumption of plant	0.110	0.890
03	X: Rayon fibre		
	Y: Normalised electrical consumption of plant	0.529	0.471
	Y: Normalised fuel oil consumption of plant	0.467	0.533
04	X: Yarn plant # 03		
	Y: Normalised electrical consumption of plant	0.151	0.849
	Y: Normalised coal consumption of plant	0.434	0.566
05	X: Synthetic fibre and yarn		
	Y: Normalised electrical consumption of plant	0.489	0.511
06	X: Viscose filled yarn		
	Y: Normalised electrical consumption of plant	0.120	0.880
	Y: Normalised fuel oil consumption of plant	0.412	0.588
07	X: Cotton yarn		
	Y: Normalised electrical consumption of plant	0.080	0.920
08	X: Fabrics plant # 01		
	Y: Normalised electrical consumption of plant	0.184	0.816
09	X: Fabrics plant # 02		
	Y: Normalised fuel oil consumption of plant	0.124	0.876
	Y: Normalised fuel oil consumption of plant	0.220	0.780
10	X: Cloth		
	Y: Normalised electrical consumption of plant	0.461	0.539
11	X: Jute goods		
	Y: Normalised electrical consumption of plant	0.832	0.168
	Average value (electrical consumption)	**0.334**	**0.666**
	Average value (fuel consumption)	**0.331**	**0.669**

Table 15: Constants in the curve fits for normalised energy consumption (NEC) for food products

Sl. No.	Food products Range of X= 0 to 1; Range of Y: 0 to 1	b_0	b_1
01	X: Bread		
	Y: Normalised electrical consumption of plant	0.258	0.742
02	X: Sugar		
	Y: Normalised electrical consumption of plant	0.176	0.824
03	X: Edible oil plant # 01		
	Y: Normalised electrical consumption of plant	0.106	0.894
04	X: Edible oil plant # 02		
	Y: Normalised electrical consumption of plant	0.112	0.888
05	X: Milk food		
	Y: Normalised electrical consumption of plant	0.431	0.569
06	X: Milk products		
	Y: Normalised electrical consumption of plant	0.538	0.462
07	X: Diary products		
	Y: Normalised electrical consumption of plant	0.671	0.329
08	X: Confectionery and food drinks		
	Y: Normalised electrical consumption of plant	0.214	0.786
09	X: Ice cream		
	Y: Normalised fuel oil consumption of plant	0.446	0.554
10	X: Fruit pulp		
	Y: Normalised electrical consumption of plant	0.412	0.588
	Y: Normalised fuel oil consumption of plant	0.576	0.424
	Average value (electrical consumption)	**0.324**	**0.676**
	Average value (fuel consumption)	**0.511**	**0.489**

Table 16: Constants in the curve fits for normalised energy consumption (NEC) for miscellaneous products

Sl. No.	Miscellaneous Range of X= 0 to 1; Range of Y: 0 to 1	b_0	b_1
01	X: Automobile tyre		
	Y: Normalised electrical consumption of plant	0.419	0.581
	Y: Normalised fuel oil consumption of plant	0.108	0.892
02	X: Kaolin		
	Y: Normalised electrical consumption of plant	0.389	0.611
	Y: Normalised fuel oil consumption of plant	0.786	0.214
03	X: Bulk drugs		
	Y: Normalised electrical consumption of plant	0.415	0.585
04	X: Toilet soap		
	Y: Normalised electrical consumption of plant	0.455	0.545
	Y: Normalised coal consumption of plant	0.790	0.210
05	X: Detergent		
	Y: Normalised electrical consumption of plant	0.389	0.611
06	X: Transparent polyvinyl chloride bags		
	Y: Normalised electrical consumption of plant	0.732	0.268
	Y: Normalised fuel oil consumption of plant	0.469	0.531
07	X: Condoms		
	Y: Normalised electrical consumption of plant	0.513	0.487
	Y: Normalised fuel oil consumption of plant	0.876	0.124
08	X: Ceramic crockery		
	Y: Normalised electrical consumption of plant	0.463	0.537
09	X: Super refractory		
	Y: Normalised electrical consumption of plant	0.517	0.483
10	X: Insulator		
	Y: Normalised electrical consumption of plant	0.170	0.830
	Y: Normalised fuel oil consumption of plant	0.403	0.597
	Average value (electrical consumption)	**0.446**	**0.554**
	Average value (fuel consumption)	**0.572**	**0.428**

Table 17 gives the SECR for the above industries. On the demand side, the increased CO_2 production is directly proportional to the SECR. It can be seen that the SECR is as high as 3.8 in certain areas.

Table 17: SEC ratio for the products

Sl. No.	Product	SEC ratio	
		Electric power	Fuel
01	Styrene	2.772	NA
02	Polystyrene	1.223	NA
03	PVC resins and compounds	1.049	1.274
04	Linear alkyl benzene	1.191	NA
05	High/low density polyethylene	2.210	1.475
06	Caustic soda	1.152	--
07	Sodium silicate	1.192	--
08	White cement	1.512	NA
09	Portland cement plant #1	1.124	1.499
10	Portland cement plant #2	1.115	--
11	Portland cement plant #3	1.584	--
12	Portland cement plant #4	1.332	--
13	Portland cement plant #5	1.381	--
14	Portland cement plant #6	1.616	--
15	Paper plant #1	1.085	--
16	Paper plant #2	1.330	--
17	Printing paper	1.988	3.423
18	Cast coated paper	2.593	1.739
19	Paper board	2.114	NA
20	Sponge iron	1.083	--
21	Cold rolled strips	2.911	NA
22	Carbon alloy steel	1.261	--
23	Steel	1.054	--
24	Steel coils	1.398	--
25	Continuous automatic castings	3.636	NA
26	Yarn plant #1	1.533	2.448
27	Yarn plant #2	1.018	--
28	Rayon fibre	2.057	NA

NA : Not available or could not be estimated

Table 17: SEC Ratio for the products (contd.)

Sl. No.	Product	SEC ratio	
		Electric power	Fuel
29	Yarn plant #3	1.065	--
30	Synthetic fibre and yarn	1.604	--
31	Viscose filled yarn	1.009	2.927
32	Cotton yarn	1.093	NA
33	Fabrics plant #1	1.213	NA
34	Fabrics plant #2	1.059	1.534
35	Cloth	1.950	NA
36	Jute goods	2.092	NA
37	Bread	2.128	1.383
38	Sugar	1.821	NA
39	Edible oil plant #1	1.035	1.035
40	Edible oil plant #1	1.110	NA
41	Milk food	1.374	NA
42	Milk products	2.281	NA
43	Diary products	2.133	--
44	Confectionery and food drinks	1.214	1.444
45	Ice cream	1.059	--
46	Fruit pulp	2.280	2.860
47	Automobile tyre	1.838	1.196
48	Kaolin	1.845	1.001
49	Bulk drugs	1.057	NA
50	Toilet soap	1.891	2.587
51	Detergent	3.801	--
52	Transparent polyvinyl chloride bags	2.858	2.524
53	Condoms	3.343	3.062
54	Ceramic crockery	1.009	--
55	Super refractory	1.873	NA
56	Insulator	0.532	8.370

NA : Not available or could not be estimated

CONTROL OF SPECIFIC ENERGY CONSUMPTION

Control of SEC is possible in two ways:

- Increasing the instantaneous production level to a point where the SEC is minimum.
- Minimising the SEC and profiling the NEC curves through standard and additional DSM strategies.

Increasing Instantaneous Production Rate

Increase of the instantaneous production rate is by:

- Maintaining the instantaneous production level at a high point corresponding to the minimum feasible SEC.
- If market considerations do not permit such a high production rate, then a production scheduling with alternate periods of high production rate and no production similar to on-off control is suggested. At present, most industries go in for uniform production rates based on the market demand. The production on-production off cycle time and sequence will be determined by the controls available with the utilities. The following modifications to the system aid in designing an optimal production on-production off sequence:
  - Sizing of equipment: The response of the energy efficiency of the utilities and end use equipment to loading determines the SEC characteristics of the product. By matching the equipment size to the production requirement, the SEC can be lowered. Based on the past history of the industrial production, a fair idea of the production levels can be obtained. This can be used as the basis for determining the equipment size. Oversized equipment often results in poor part load efficiencies (and hence high SEC) at low loading levels.
  - Matching of equipment: In multiple operation processes, matching of equipment size from one process to another can result in reduced energy consumption. Situations where there is excessive capacity in one process and bottleneck (large waiting time) in the next can reduce the overall production rate, thereby increasing the constant component of energy consumption and consequently the SEC.
  - Tracking of the load by the source: In a large number of industrial systems there is no communication between the utility and end use equipment. The various methods of thermal and mechanical tracking of utilities and end use equipment are on-off controls, proportional integral differential controls or logic based controls (fuzzy logic, etc.). These can be analog controls or direct digital controls through remote operation.
  - Review of constant energy equipment: Equipment where energy consumption is independent of the production level such as lighting, room air conditioning, etc., need to be reviewed and minimised through the use of more energy efficient equipment and the use of effective controls for modulating energy consumption.

Direct digital control with logic base for lighting, heating, ventilation and air conditioning can bring about drastic reduction in the constant energy consumption.

- A high instantaneous production level cannot be maintained in some cases, due to frequent outages and breakdowns,. Under such conditions, the maintenance practices must be strengthened and frequently failing components critical to production must be identified and replaced.

Minimising SEC through Standard and Additional DSM Strategies

A detailed survey of demand side management (DSM) options, awareness, barriers and implementability has been undertaken by Parikh et al. (1996). The standard DSM solutions have been explored and the following problems have been encountered in the implementation stage in this study:

- **Energy efficient motors:** They are costlier (capital cost) by 20-25 % as compared to standard motors. Under harsh and poor quality electrical environment, they are prone to winding failures. Rewound motors no longer provide the original energy efficiency.
- **Variable speed drives and heat driven refrigeration systems**: The technology is still in the developmental phase and considerable work is required before the problem of reliability, harmonics and high capital cost are solved.

The other DSM options such as combined heat and power, use of lighting controls, power factor improvement, etc., are highly successful in reducing the SEC.

Besides the standard DSM options, some of the measures which have been successfully implemented for control of SEC are described below:

- Minimising direct dispersion of energy by reducing the controllable losses.
- Operational optimisation: This provides a simple low cost solution to reducing the SEC in utilities as well as end use equipment.
- Retrofitting by energy savers, material/energy feed back loops, recuperators, etc: Recycling or recuperation of energy being wasted into the atmosphere, reduce both the energy and environmental burdens.
- Integration of heating, cooling and power generating equipment (Bhatt et al., 1998).
- Substituting by low energy processes: Replacement of entire sub-systems by systems using less energy intensive processes such as evaporative coolers, steam jet ejector will reduce the overall SEC.

By implementing the above solutions, it has been possible to successfully reduce the no load energy consumption and overall SEC. Most of the problems of high SEC in the industries studied, fall under the above categories. These have been suggested, implemented in many industries and visible reduction in SEC has been established.

CONCLUSIONS

The major conclusions of the study is that:

1. The novel and primary DSM technique proposed herein is to maximise the instantaneous production rate to a point where the SEC and CO_2 emission levels are only 30 % of their no load levels.

2. If the market demand does not permit such a high production level, then a strategy of production on-production off must be adopted (maintaining full production during the period of production on). An optimal production-on-production off sequence can be worked out for each industry considering the product demand as well as the nature of the utilities.

3. For optimal production on-production off sequencing, the following modifications in the system are suggested:
   - Matching of the capacity of equipment.
   - Minimising energy consumption of equipment which have no bearing on the production rate.
   - Introduction of communication between the utilities and end use equipment.
   - Installation of efficient controls for utilities.

4. A design for improved SEC must aim at having a value of b_0 (no load energy consumption) approaching zero and minimise b_1 (the sensitivity of energy consumption to production level). This can be achieved by reduction in the SEC of industrial products by as much as 30-40 % by certain simple measures such as:
   - Installation of energy savers.
   - Operational optimisation
   - Minimising direct energy dispersion
   - Substituting by low energy processes

REFERENCES

Boustead, I. and Hancock, G.F. (1979) **Handbook of Industrial Energy Analysis**, Ellis Horwood Limited, Chichester, UK.

Bhatt Siddhartha M., Mandi, R.P. and Nandakumar V.N. (1998) 'Integration of energy systems in the small capacity range', *Energy Sources*, Vol.20, Nos. 4-5, pp. 407-426.

Gellings, C.W. (1997), 'Industrial electrotechnology development', *Int. J. of Global Energy Issues*, Vol. 9, Nos. 4-6, pp.256-263.

Otfried Voigt (1995), 'Power Plants', **Proc. GTDC Symposium 1995** at Trivandrum, Published by State Committee on Science and Technology, Trivandrum, India, pp.99-102.

Painuly, J.P., Bhattacharya, K. and Parikh, J. (1997) 'Joint implementation for carbon dioxide reduction in India: an analysis of auxiliary reduction in power plants', *Int. J. of Global Energy Issues*, Vol. 9, Nos. 4-6, pp.275-285.

Parikh, J.K., Reddy, B.S., Banerjee, R. and Koundinya, S. (1996) 'DSM survey in India: Awareness, barriers and implementability', *Energy*, Vol. 21, No.10, pp.955-966.

Perry, R.H. and Green, D. (1984) **Perry's Chemical Engineers' Handbook**, McGraw Hill, Singapore.

Sivaramakrishnan, R. and Siddiqi, T.A. (1997) 'Growth of electricity use in India, and strategies for reducing greenhouse gas emission', *Int. J. Global Energy Issues*, Vol.9, Nos.4-6, pp.225-236.

Nomenclature

a_0, a_1	Constants in curve fits
b_0, b_1	Constants in curve fits
C	Calorific value (kJ/kg)
CF	Capacity factor (0 to 1)
DSM	Demand side management
f	Fraction (0 to 1)
I	Index of green house gas emission
NEC	Normalised energy consumption (0 to 1)
SEC	Specific energy consumption
SECR	Specific energy consumption ratio (typically 1 to 5)
SI	Specific CO_2 generation index

Greek letters

η	Thermal efficiency (% or dimensionless)

Subscripts

A	Auxiliary power in fossil fuel power plant
E	Gross electric conversion
T	Electrical transmission and distribution

Chapter 6

ECONOMIC GROWTH, ENERGY USE, AND CO_2 EMISSIONS

Jordi Roca
Dpt. Teoria Econòmica. Universitat de Barcelona.
jroca@eco.ub
Vicent Alcántara
Dpt. Economia Aplicada. Universitat Autònoma de Barcelona.
vicent.alcantara@uab.es

1. THE ENVIRONMENTAL KUZNETS CURVE HYPOTHESIS AND CO_2 EMISSIONS

In the last 10 years, debate concerning how economic growth and environmental impact are related has mainly revolved around the so-called environmental Kuznets curve (EKC) hypothesis. This hypothesis states that at the first stage of economic development, environmental pressures increase as per capita income increases, but after a critical turning-point these pressures diminish along with higher income levels.[1] The name environmental Kuznets curve derives from the similarity with the relationship between the level of inequality and per capita income considered by Simon Kuznets (1955). The most optimistic authors writing on the matter even suggest that economic growth is itself the solution to environmental problems. In this respect, Beckerman stated: "... although economic growth usually leads to environmental degradation in the early stages of the process, in the end the best -and probably the only- way to attain a decent environment in most countries is to become rich" (Beckerman, 1992 quoted in Stern et alt, 1996, p. 1152). Nevertheless, we do not agree with this ingenuous interpretation at all.

From an empirical point of view, although there is certain evidence that some environmental indicators are better in rich countries than in average-income countries, none of the pollutants examined in the literature fulfils the EKC hypothesis unequivocally. The

[1] When the hypothesis does not directly refer to pollutant generation but to the (closely related) use of natural resources, it is sometimes known as the "dematerialization" hypothesis.

outcomes, while not totally conclusive, are generally more favourable to this hypothesis for pollutants with local and regional impacts and a relatively low cost of abatement, as is the case of sulphur atmospheric emissions.[2] In the case of global pollutants – like CO_2, the pollutant which interests us in this article – econometric estimates have almost never been favourable to the EKC hypothesis. On the contrary, per capita emissions show a highly positive correlation to per capita income even for the high income range. This is not really surprising: rich people have incentives to improve the environment or to reduce environmental risk when they themselves are those most affected by such problems, but this is not the case when risks are mainly transferred across a great distance and/or to a remote future (Ansuategi et al., 1998). In such cases, the people with high consumption levels have few incentives to bring their lifestyle into question.

Some authors have suggested that the endogenous change in the demand structure for goods and services, which would go hand in hand with economic growth, could explain the EKC. If this argument is taken into account, the sectors that become increasingly important as per capita income rises are those which have less environmental impact. The evidence that generally underlies this position is the increasing demand directed at the service sectors at the expense of demand directed at the industrial sector. However, much more empirical research needs to be done on the assumptions this argument is based on: some activities that are included in the services sector may have as much or more environmental impact (direct and/or indirect) as some industrial activities (consider, for example, long-distance tourism). Furthermore, the observed EKCs could result (at least partly) from a displacement of the most polluting industries from the rich countries to the poorest ones, without an ensuing substantial variation in the composition of consumption (and its pollution content) (Arrow et al, 1995; Stern et al, 1996; Ekins, 1997; Muradian and Martínez-Alier, 2001).

In any case, the argument relating to demand composition would at most explain a reduction in environmental pressures *per unit of GDP* as income increases: it would not account for a reduction of these pressures in absolute terms. Let us consider this in greater detail.

Let's take P to be an indicator of environmental pressure, such as emissions of CO_2. We can write P = (P/GDP)*(GDP). Following Bruyn and Opschoor (1997), we can say that there exists "delinking" between environmental pressure and economic growth in its strong sense if:

$$dP/dt < 0$$

That is to say, if

$$[(d(P/GDP)/dt) * GDP + (P/GDP) * d(GDP)/dt] < 0$$

[2] Some of the first studies were by Grossman and Krueger (1991), Shafik and Bandyopadhyay (1992), Panayotou (1993), Selden and Song (1994), and Holtz-Eakin and Selden (1995). Special issues of *Ecological Economics* (1995, 1998) and *Environment and Development Economics* (1996) have discussed the EKC hypothesis. Ekins (1997), Stern et al (1996) and Stern (1998) make critical reviews of the literature. Most studies have been performed with panel or cross section data from a series of countries (or from different US states). We think it is more appropriate to study the relationship between economic growth and each type of environmental pressure, analysing the experience of individual countries and using both econometric and historical analysis (Stern et al, 1996). Some studies which adopt a longitudinal analysis are: Bruyn et al (1998) and Roca et alt (2001), about various atmospheric pollutants for four developed countries and for Spain respectively. The

which is the same that

$$- [(d(P/GDP)/dt) /(P/GDP)] > d(GDP)/dt / GDP$$

This means that economic growth would only be environmentally "friendly" if the intensity of environmental pressure lessened *due to* the increase in GDP and, moreover, the rate of decrease of P/GDP ("weak delinking") were higher than the rate of GDP growth.

The distinction between "delinking" of economic growth and environmental pressures in the "weak (or relative) sense" and in the "strong (or absolute) sense" must be highlighted because it occasionally gives rise to some confusion. By way of example, Sun (1999) discusses the hypothesis of the environmental Kuznets curve for CO_2 emissions. He defines this hypothesis as a performance of the intensity of emissions (relationship between emissions and GDP) that is an inverted-U shaped function of the per capita GDP. According to this author, the hypothesis would actually be fulfilled in some cases (he analyses the case of China) although he wisely indicates that the result should not be generalised for all countries, and that the conclusions should not be extended to any type of environmental indicator. However, the decrease in the relationship between CO_2 and GDP of China, which began years ago, has not been accompanied by any cutback in the absolute level of emissions (neither total or per capita emissions) at all. Another example of the confusion that could be provoked is, in our opinion, to be found in the chapter on "Decarbonizing the Energy Economy" (Dunn, 2001) published in the prestigious report of the Worldwatch Institute for the year 2001. Both the data relating to the improvement in the relationship between carbon emissions and total production in most rich countries (as well as other economies like China) and the historical transition from solid to liquid fuels and from liquid to gas fuels are considered in this report as indicating that the desired "decarbonization" process has been under way for a long time. However, it does not state clearly that the environmentally relevant data is the evolution of the total CO_2 emissions. This is a very politically relevant distinction if we take into account the Bush Administration's refusal to ratify the Kyoto Protocol on climatic change, which committed the USA to reducing 1990 levels of emissions by 7% in the period 2008-2012, and its alternative proposal to reduce the intensity of emissions in relation to GDP.

2. Economic Activity and CO_2 Emission Relationship: Determining Factors

As seen, the evolution of CO2 emissions may result from two factors: an indicator of economic activity (as the GDP measured at constant prices is, that is to say, as if prices didn't change) and an indicator of emission intensity (CO_2/GDP). [3] In the case of the latter, which

conclusions from longitudinal analysis are generally even more sceptical about the inverted U-shaped EKC than from panel data.

[3] This numerical separation does not imply a separation into explanatory factors since they do not evolve independently of each other. For example, in energy economics it has been discussed that the very increase in energy efficiency could stimulate energy consumption. For instance, if vehicles are more efficient and run on less petrol, this will make it cheaper to travel by car and may encourage people to use the car and thereby consume more petrol. This effect is sometimes known as the "Jevons paradox" (Giampietro, 1999) because the famous English economist alluded to the subject in his book *The Coal Question* (Jevons, 1865). It has also

measures the existence or absence of "weak (or relative) delinking" as defined above, there has recently been some controversy surrounding its determining factors. More specifically, the relative importance of, on the one hand, energy intensity and, on the other, the "carbonization index" has been debated. Thus, Mielnik and Goldemberg (1999) question the main role that, according to them, has been given to energy intensity (defined as the relationship between primary energy and GDP). In their opinion, the evolution of energy intensity has a well-defined historical pattern while the second factor, the carbonization index (defined as the relationship between the emissions of CO_2 and the primary energy), is much more variable. Ang (1999), in response to the previous comment, insists on the greater importance of energy intensity.

In our opinion, the general debate about whether the main factor in the evolution of emission intensity (CO_2/GDP) is energy intensity or the carbonization index sheds little light on the matter.

The general relationship is[4]

$$CO_2/GDP = (CO_2/E)*(E/GDP)$$

where E represents the use of primary energy.

Whether the changes in one factor or the other are more important is a purely empirical issue that, however, can not be solved once and for all from the experience in various countries or groups of countries during a given period of time. There are reasons -and empirical evidence- for thinking that both factors can be very variable. Thus, energy intensity is an aggregate that depends on several factors, among which the structure of goods and services demand, energy efficiency in the provision of different goods and services and transport models can be highlighted. It seems clear that energy intensity can vary considerably, even among countries with a similar level of per capita GDP, as the comparison, for example, between the United States and Japan shows. As regards the carbonization index, it is important to see that the different energy sources imply an extremely different CO_2/energy relationship: for example, in the case of solid fossil fuels, the carbon emission factor is slightly higher than for natural gas, and the value is null (directly) for wind power and solar energy as well as nuclear energy.

In the following empirical analysis, we will see the relative role both factors play in explaining, firstly, the differences in the relationship between emissions and GDP in different regions of the world for 1999; and secondly, the changes that took place between 1973 and 1999, which means we simply update Ang's analysis (1999), taking the current database of

been called the *rebound effect* (see *Energy Policy* -vol. 28, n. 6-7- edited by L. Schipper, dedicated to it). In extreme cases, the very increase in efficiency could be counterproductive from the point of view of reducing demand, as Jevons' general conclusion seemed to be. However, we should distinguish between two facts. The first one is that the effect actually exists and the second one, whether it is sufficient to counteract the positive effects of an improvement in efficiency. The effect undoubtedly exists but in most cases it is highly unlikely to totally counteract the positive effect of the improvement in efficiency.

[4] The CO_2 emissions considered in this article are only the emissions related to fossil fuel combustion. These emissions represent the immense majority of total emissions. They do not include emissions from some industrial processes or those from biomass burning. In this latter case, emissions should only be counted when biomass is exploited on a unsustainable basis since if the exploitation is sustainable, CO2 emissions are compensated for by later carbon storage as the biomass grows, that is the emissions are more or less null in net terms.

the IEA as our source[5]. Both GDP and energy data must not be considered completely reliable as indicators of economic activity or of the exosomatic energy needed to perform this activity. There are at least two reasons for this. In the first place, GDP is known to disregard most activities that do not form part of the market and are not financed by the public sector; in the second place, both economic and energy data are of limited reliability, especially where poor countries are concerned. Even if these limitations are taken into consideration, the analysis is of great interest.

Table 1 shows data related to the two abovementioned factors, which determine the relationship between CO_2 and GDP emissions (in constant dollars and purchasing power parities) for the whole world divided into regions, while table 2 shows how these factors have evolved in the last few decades for the same regions. An analysis of the data allows us to conclude, as Ang (1999) did, that the current differences can be given a clearer explanation using the energy intensity factor and that, in almost all the regions, the relative changes in energy intensity were also much higher than changes in the carbonization index. Values for energy intensity in 1999 range between a minimum of 180.9 tonnes of CO_2 per million US dollars in 1995 (for Latin America) and a maximum of 614.3 tonnes (for the former USSR), while CO_2 tonnes generated by every primary energy unit equivalent to a ton of oil varied between 1.5 (for Africa) and 2.8 (for China) (table 1). As regards historical trends between 1973 and 1999, energy intensity changes vary from the spectacular drop of 69.45% (China) to an increase of 184.71% (Middle East) (table 2), while the carbonization index shows a decrease of 23.76% in European OECD countries in comparison to an increase of 48.41% in the case of Asian countries (excluding China).[6]

However, this interesting empirical evidence should not induce us to think that the carbonization index is of little importance. This is particularly true if we think that the decisions about environmental policies can have a big influence on this factor. Thus, if a high carbon tax were approved, as has been proposed many times in Europe (Weisäcker and Jesinghaus, 1992), the substitution between energy sources could be relevant, as it would reduce this index at the same time as energy intensity would also diminish in response to the change in the relative price of energy. Which of these two factors would be more influenced is difficult to foresee and it depends much on other factors (for example, on how investment in public transport evolves or what subsidies are given to alternative energy sources). Ang is completely right when he says: "changes in the energy intensity can be caused by a greater number of factors as compared to the carbon factor" (Ang, 1999, p. 944), but it is also important to remember that while it is impossible to imagine a future economic and social system without energy use, it is, however, quite possible to imagine a future in which a complete transition from fossil fuels to other energy sources has occurred. Moreover, such a future is unavoidable in the long term.

---

5 The current energy database includes estimates on biomass use, unlike previous years when such data were excluded. This is the main reason (apart from considering different period in time) why our results differ from Ang's (1999).

6 In a previous paper on a relatively much more homogeneous territory (the 15 European Union countries), we showed that the variability in the carbonization index (where the maximum was in Greece and the minimum in Sweden and France) is very similar and even somewhat bigger than the variability in energy intensity (Padilla y Roca, 2001).

Table 1. Carbon intensity, carbonization index and energy intensity for World regions, 1999

	CO2/GDP(1)	CO2/E(2)	E/GDP(3)
OECD Europe	417.9	2.2	187.7
USA and Canada	649.1	2.4	268.4
OECD Pacific	456.2	2.3	195.4
Non-OECD Europe	716.5	2.5	284.9
Former USSR	1530.5	2.5	614.3
China	690.0	2.8	249.7
Asia (excluding China)	429.1	2.0	217.6
Latin America	371.8	2.1	180.9
Middle East	961.8	2.6	372.6
Africa	492.0	1.5	330.1
World	581.6	2.4	246.9

Units:

(1): Tonnes CO_2 per Million 1995 US$ in power purchasing parities.

(2) Tonnes CO_2 per Tonnes of oil equivalent primary energy

(3) Energy intensity in Tonnes of oil equivalent primary energy per Million 1995 US$ in power purchasing parities.

Source: Calculated from IEA (2001), *CO_2 emissions from fuel combustion.* OECD/IEA

If we look at table 2, we see that there was a great reduction in the relationship between emissions and total GDP in the whole world, but owing to the growth in global economic activity, total emissions increased more than 40%. Using the terms of the previous section, we could speak of "relative delinking" but not of "absolute delinking" at all.

Table 2. Changes in carbon intensity, carbonization index and energy intensity for World regions, 1973-1999

	% total change		
	CO2/GDP	**CO2/E**	**E/GDP**
OECD Europe	-45.66	-23.76	-28.73
USA and Canada	-43.97	-9.11	-38.36
OECD Pacific	-30.14	-17.62	-15.21
Non-OECD Europe	-40.32	-12.62	-31.70
Former USSR	4.61	-16.33	25.03
China	-62.29	23.45	-69.45
Asia (excluding China)	10.96	48.41	-25.24
Latin America	2.97	3.46	-0.47
Middle East	182.55	-0.76	184.71
Africa	36.02	17.08	16.17
World	-35.03	-10.39	-27.50

Source: Calculated from IEA (2001), *CO_2 emissions from fuel combustion.* OECD/IEA

3. A Particular Case: The Evolution of CO_2 Emissions in Spain

In this section we will give a more detailed analysis, over a longer period of time, of the experience of one particular OECD country, namely Spain, from 1964 to 1999. A specific point of interest offered by this case is, as we shall see, that it shows how the improvement of an environmental pressure indicator (in this case the "weak or relative delinking" indicator between CO_2 and GDP during the 1980s) can be both accompanied and explained by the worsening of another indicator, namely electricity generation using nuclear power. Therefore, we need to be cautious about reaching conclusions regarding the relationship between economic growth and environmental pressure, not only because we must consider the distinction between reductions in environmental pressures in absolute and relative terms, but also because it is necessary to visualise the whole, bearing in mind the kinds of interrelationship that exist between different types of environmental pressures.

Figure 1 shows the evolution of the three relationships analysed in the second section. According to these figures, in Spain –unlike the pattern observed in most industrialised countries and, in fact, in the whole world (table 2)- energy intensity has not shown a tendency to decrease in the last few decades. From the mid-60s until 1981 it grew approximately 60%, that is, the use of primary energy increased much faster than growth in GDP, and it was more or less stable in the next two decades, although it reached historic maximum values in the second half of the 1990s. Energy intensity is an aggregated statistical relation dependent on how efficiently the energy is used for different purposes but it also depends -as already pointed out- on many other factors. In Spain there is no doubt that until 1999 the main factor which prevented any improvement in the global evolution of this indicator -in spite of technological efficiency improvements in the use of energy- was the increasing use of fuel in road transport (Alcántara and Roca, 1995). At the end of the 1990s, energy consumption for road transport was 31% of the total amount of final energy demand, whereas in the mid-60s it was only 16% of the total.

Figure 1. Evolution of CO_2 emissions per unit of real GDP, energy intensity and carbonization index, Spain, 1964-1999

Index number: 1964=1

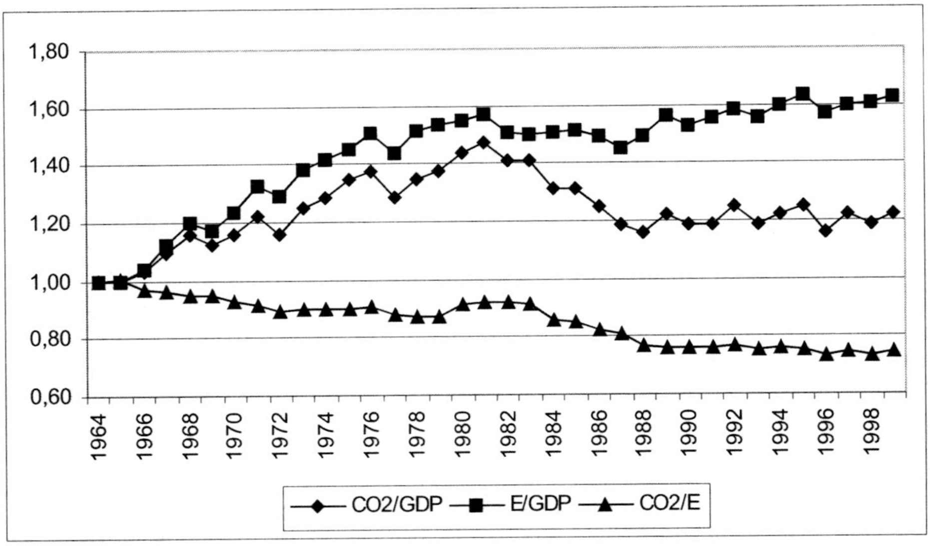

Source: Calculated from IEA (2001), *CO_2 emissions from fuel combustion*. OECD/IEA

As regards the "carbonization index", significant changes can be observed during two periods. The first change took place between 1965 and 1972 when there was an overall reduction of more than 10%; the second change took place during the 1980s. In both cases, what underlies are changes in the structure of primary energy sources. The changes are complex but the decreasing importance of coal predominated in the first change, while the second one was basically associated with the increasing use of nuclear energy.

As a whole, the relationship between CO_2 emissions and GDP went through several phases and was 20% higher in 1999 than in 1964 as a result of two significant variations: an increase of approximately 60% in energy intensity and a decrease of practically 30% in the relationship between CO_2 and primary energy (figure 1). Figure 2 shows how this indicator evolved attending the changes in per capita income. The EKC hypothesis is clearly not fulfilled, not even in a "relative" or "weak" sense. The Spanish economy only could seem to have entered in the decreasing phase of the hypothetical curve between 1981 and 1988 and this was due, unfortunately, to the greater use of nuclear energy rather than a reorientation to the use of clean energy sources.

Figure 2. CO_2 emissions per unit of GDP, Spain, 1964-99

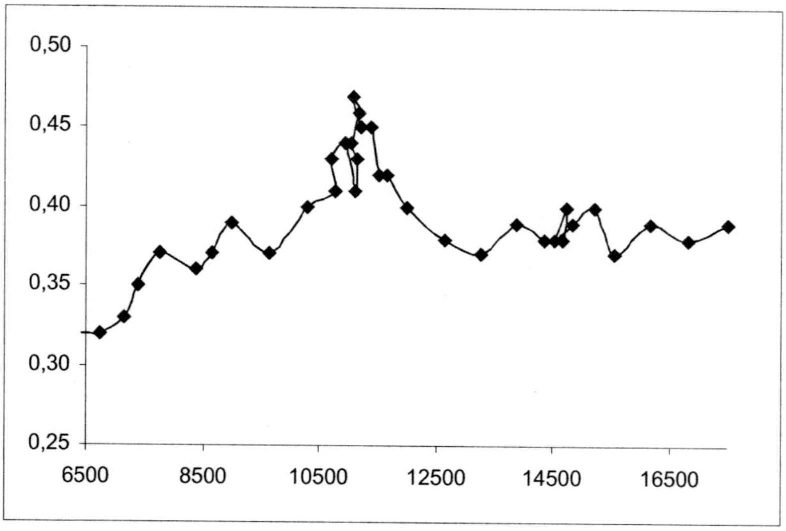

Axis y: Tonnes divided by 1,000 1995 US$ in power purchasing parities.
Axis x: Per capita GDP in 1995 US$ in power purchasing parities.
Source: Calculated from IEA (2001), *CO_2 emissions from fuel combustion*. OECD/IEA

As regards the evolution of primary energy use and emissions in per capita terms, figures 3 and 4 show how both magnitudes grew almost continuously and considerably as per capita income grew. Even in the 1980s, with relative delinking between economic growth and CO_2 emissions, these emissions failed to decrease in absolute terms.

Figure 3. Per capita primary energy, Spain, 1964-99

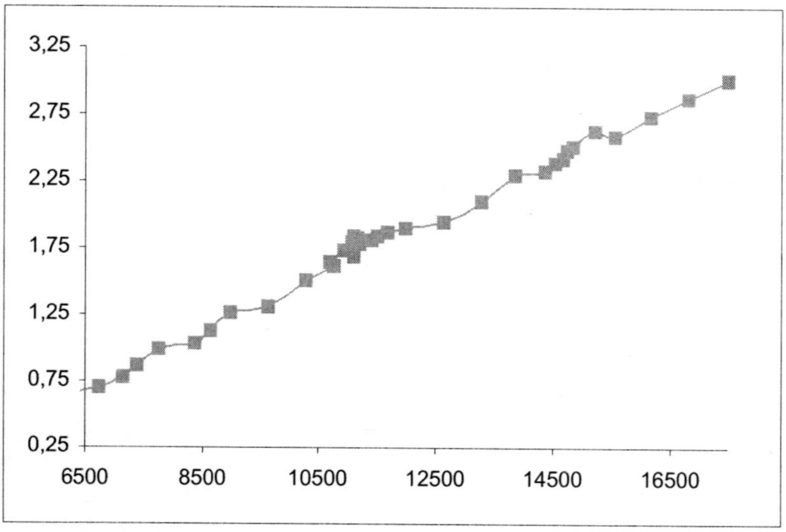

Axis y: Tonnes oil equivalent
Axis x: Per capita GDP in 1995 US$ in power purchasing parities.
Source: Calculated from IEA (2001), *CO_2 emissions from fuel combustion*. OECD/IEA

Figure 4. CO_2 per capita emissions in relation to the level of per capita GDP, Spain, 1964-99

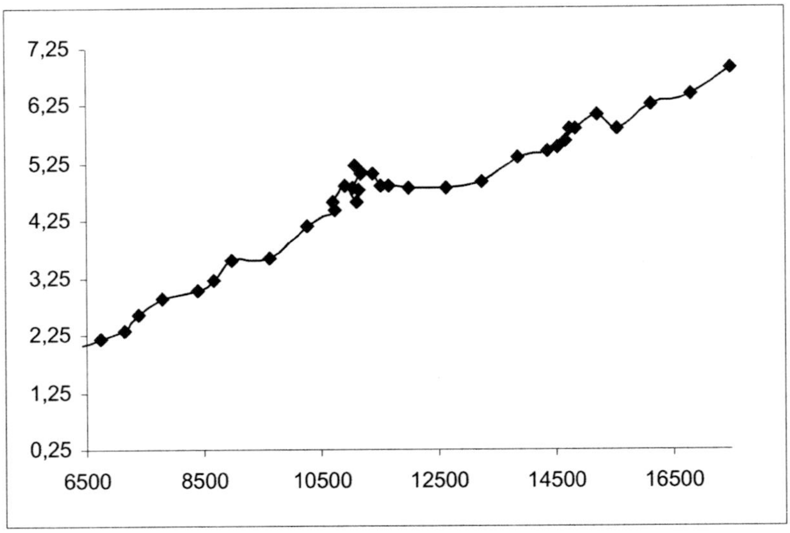

Axis y: Tonnes

Axis x: Per capita GDP in 1995 US$ in power purchasing parities

Source: Calculated from IEA (2001), *CO_2 emissions from fuel combustion*. OECD/IEA

4. Final Comments

Although most economists have been wary about reaching definitive conclusions, they have sometimes defended the hypothesis that economic growth itself would bring about changes that would improve the quality of the environment. Without denying that economic growth may have favoured a solution to some specific environmental problems in certain periods in some countries, our own conclusion is that this will probably not occur in the case of CO_2 emissions. To support our discussion we have brought forward factors that determine this evolution. In the specific case of Spain, a rich economy that entered a "tertiarisation" process decades ago, we have seen that this cannot be said to have entered a "decarbonization" stage even in the limited sense of a reduction in CO_2 emissions per GDP unit.

However, there are many ways which would allow Spain (and other countries) to begin a new stage. Firstly, priority must be placed, as in most rich countries, on slowing down the increase in energy consumption for transport. Secondly, the change to long term sustainable energy sources, which would not bring about the problems associated with fossil fuels (nor with nuclear power) must be promoted. Finally, there must be a change in the orientation of economic policies. Until now economic growth is generally the lead indicator for economic policy but we think that it is becoming more and more clear that economic growth, at least in rich countries, is neither a necessary nor a sufficient condition for improving human well-being.

REFERENCES

Alcántara, V. and Roca, J., (1995), Energy and CO_2 emissions in Spain, *Energy Economics*, 17 (3), 221-230.

Ang, B. W., (1999), Is the energy intensity a less useful indicator than the carbon factor in the study of climate change?, *Energy Policy*, 27, 943-946.

Ansuategi, A., Barbier, E. and Perrings, Ch., (1998). The Environmental Kuznets Curve. In: van den Bergh, J.C.J.M., Hofkes, M.W., *Theory and Implementation of Economic Models for Sustainable Development*. Kluwer Academic Publishers.

Arrow, K., Boling, B., Costanza, R., Dasgupta, P., Folke, C., Holling, S., Jansson, B.-O, Levin, S., Mäler, K.-G., Perrings, C. and Pimentel, D., (1995), Economic growth, carrying capacity and the environment. *Science* 268, 520-521.

Beckerman, W., (1992), Economic growth and the environment: Whose growth? *World Development* 20, 481-496.

Bruyn, S. M. de and Opschoor, J. B., (1997), Developments in the throughput-income relationship: theoretical and empirical observations, *Ecological Economics*, 20, 255-268.

Bruyn, S.M. de, Bergh, J.C.J.M. van den and Opschoor, J.B. (1998), Economic growth and emissions: reconsidering the empirical basis of environmental Kuznets curves, *Ecological Economics*, 25, 161-175.

Dunn, S., (2001), Decarbonizing the Energy Economy. In: Brown, L., and alt. *State of the world*. Worldwatch Institute.

Ekins, P., (2000), *Economic Growth and Environmental Sustainability. The prospects for green growth.* Routledge, London and New York.

Giampietro, M., (1999), Economic growth, human disturbance to ecological systems, and sustainability. In: Walker, L. R. (ed), *Ecosystems of Disturbed Ground*, Elsevier.

Grossman, G.M. and Krueger, A.B., (1991), *Environmental Impacts of a North American Free Trade Agreement.* NBER Working Paper 3914, National Bureau of Economic Research (NBER), Cambridge.

Holtz-Eakin, D. and Selden, T.M., (1995), Stoking the fires? CO_2 emissions and economic growth. *Journal of Public Economics* 57, 85-101.

International Energy Agency, (2001), *CO_2 emissions from fuel combustion,* OECD, Paris.

Jevons, W.S., (1865), *The Coal Question.* MacMillan London and Cambridge.

Kuznets, S., (1955), Economic growth and income inequality. *American Economic Review* 45, 1-28..

Mielnik, O. and Goldemberg, J., (1999), The evolution of the "carbonization index" in developing countries, *Energy Policy*, 27, 307-308.

Muradian, R. and Martinez-Alier, J., (2001). Trade and environment: from a 'Southern' perspective. *Ecological Economics* 36, 281-297.

Padilla, E. and Roca, J., (2002), The proposals for a European tax on CO2 and their implication for intercountry distribution. Presented at 7th Biennal *Conference of the International Society for Ecological Economics.* March 2002. Tunisia.

Panayotou, T., (1993), Empirical Tests and Policy Analysis of Environmental Degradation at Different Stages of Economic Development. Working Paper, Technology and Environment Program, International Labour Office, Geneva.

Roca, J., Padilla, E., Farré, M. and Galletto, V., (2001), Economic growth and atmospheric pollution in Spain: discussing the environmental Kuznets curve hypothesis. *Ecological Economics* 39, 85-99.

Selden, T.M. and Song, D., (1994), Environmental quality and development: is there a Kuznets curve for air pollution estimates? *Journal of Environmental Economics and Management* 27, 147-162.

Shafik, N. and Bandyopadhyay, S., (1992), Economic Growth and Environmental Quality: Time Series and Cross-Country Evidence. Background Paper for World Development Report 1992, World Bank, Washington, DC.

Stern, D.I., (1998), Progress on the environmental Kuznets curve? *Environment and Development Economics* 3, 173-196.

Stern, D.I., Common, M.S. and Barbier, E.B., (1996), Economic growth and environmental degradation: the environmental Kuznets curve and sustainable development. *World Development* 24, 1151-1160.

Stern, D.J., Common, M.S. and Barbier, E.B., (1996), Economic growth, trade and the environment: implications for the environmental Kuznets curve. *World Development* 24, 1151-1160.

Sun, J. W., (1999), The nature of CO_2 emissions Kuznets curve, *Energy Policy,* 27, 691-694.

Weizsäcker, E.U. von and Jesinghaus, J., (1992), *Ecological Tax Reform*, Zed Books, London

Acknowledgements: We would like for the financial support of the PB98-0868, BEC2000-415 and SGR2001-161.

Chapter 7

RISK MEASURES AND THE COST OF EQUITY IN THE ENERGY INDUSTRY

Perry Sadorsky
Schulich School of Business
York University
4700 Keele Street
Toronto, Ontario, Canada M3J 1P3
Email: psadorsk@schulich.yorku.ca

ABSTRACT

Individual companies and industries must earn an adequate rate of return over time to continue in business. Failure to do so leads to eventual withdrawal of capital which has many consequences including contraction, re-structuring, consolidation, bankruptcy or closure. It is therefore necessary to have good measures of equity risk for managers, planners, and investors. The cost of equity is important in valuing new investment opportunities and evaluating the ongoing performance of existing business projects. This is especially true in the capital intensive energy industry. The purpose of this paper is to calculate the cost of equity in the energy industry, using several different risk measures, and to compare it with the cost of equity in other industries. The risk measures include systematic risk, total risk, downside risk, and value at risk. The empirical results indicate that while different risk measures produce different values for the cost of equity, the oil and gas industry is no more risky than many other industries. The energy equipment and services industry is, however, riskier than other industries. These results are useful for managers, budget planners, policy makers, and investors who are interested in the energy industry.

Acknowledgement: I would like to thank Irene Henriques for helpful comments

1. INTRODUCTION

The trade off between risk and return is of paramount importance to investors, portfolio managers, and capital budget planners. These individuals generally prefer investments with high return to risk characteristics. This is particularly relevant in volatile industry sectors like the energy industry where there are many sources of risk. These risks include various market risks, credit risks, and liquidity risks. Market risks arise from the possibility that changes in macroeconomic factors like currency exchange rates, interest rates, or oil and natural gas prices will adversely affect the value of an energy company's assets, liabilities, or expected future cash flow (Sadorsky (2001)).

Currency exchange risk arises whenever a company converts foreign sales and cost figures back into domestic prices. A low domestic currency can have a positive impact on a company's financial statements when foreign sales are booked back to domestic currency units. A low domestic currency can have a negative impact on a company's financial position if a domestic company imports a lot of machinery and equipment. A high domestic currency has a similar but opposite affect making imported goods and services cheaper to buy but making exports less competitively priced. Currency exchange rate risk can be managed through the use of financial derivatives like hedging, futures, and forward contracts.

The energy industry is very capital intensive. It can easily cost billions of dollars to develop a new oil field or build a new electricity generation plant. In most cases the money to finance such a project needs to be raised either through borrowing or through issuing equity. The ability to raise money in the equity markets depends upon the strength of the company's financial position (in particular its stock price) and the opportunity cost of other investment projects. Well performing companies usually have easier access to the financial markets. An energy company is exposed to interest rate risk on short and long term floating rate financial instruments as well as the refinancing of fixed rate debt. Financial risk management usually tries to achieve a balance between low cost high risk floating rate debt and high cost low risk fixed rate debt. Many energy company borrowings are in U.S. dollar denominated debts and, as a result, also exposed to currency exchange rate risk.

Most oil companies are price takers in the global market for crude oil but energy prices like the prices of oil, natural gas, and electricity tend to be very volatile and this can create large fluctuations in energy company profits. Higher energy prices generate higher profits which can, in turn, be used to increase capital expenditure budgets. Low energy prices depress profits and cash flows and generally imply weaker future capital spending. For the Canadian oil and gas industry, market betas are approximately 0.71 and oil betas are 0.31 (Sadorsky (2001)). This means that a 1% increase in stock market returns increases the stock return of Canadian oil and gas companies by 0.71%. A 1% increase in oil prices increases the stock return of Canadian oil and gas companies by 0.31%. Energy prices tend to be very volatile. For example the annualized price volatility of crude oil is approximately 25% and the annualized price volatility of natural gas is approximately 40% (Fusaro (1998)). Crude oil prices are very difficult to predict with accuracy but some progress has been made with recent econometric techniques (Sadorsky (2002)). Energy price risk can be managed using financial derivatives like options, futures, and swaps.

Credit risk arises from the possibility that one of the parties involved in a financial transaction defaults on their obligations (Fusaro (1998)). Companies can manage this type of risk by only entering into contracts with high credit rated parties.

Liquidity risk arises from the possibility of a particular company's credit rating being lowered. In such a case, it may be difficult for the company to get external funding. Liquidity risk can be managed by ensuring a company remains competitive and profitable. Credit risk and liquidity risk are company specific while market risk is not. This paper focuses on the relationship between market risks and the expected rates of return.

The exploration and production side of the energy industry tends to be very cyclical and follow the overall business cycle closely. This means that when the economy is expanding and commodity prices are strong, cash flows are high. When commodity prices weaken, cash flow quickly declines. This creates an investment opportunity for the short term investor. The cyclical nature of the industry creates cyclical investment. Investors buy exploration and production (E&P) stocks as oil prices move upwards and sell E&P stocks as oil prices peak or weaken. At the peak of the equity cycle E&P companies often spend their cash flow on acquisitions rather than on wealth creation. Buying other companies at the top of the equity cycle means that there is often overpayment for bought companies. The problem is that oil companies have a history of wasting a lot of money at the top of the commodity cycle. When oil companies make large profits, this often leads to an expansion acquisition strategy whereby large amounts of money are paid out to increase the size of the firm but not necessarily increase the return on investment (Stern Stewart (2001)). Recently there has been a lot of merger activity in the oil and gas industry. The merger between Exxon and Mobile, the merger between BP, Amoco, and Atlanta Richfield, and the merger between Elf and Total Fina illustrate the current trend towards bigger oil companies getting even bigger. But there is evidence to support the idea that while mergers increase the size of firms, size tends to increase complexity rather than savings (Ghemawat and Ghadar (2000)).

Individual companies and industries must earn an adequate rate of return over time to continue in business. Failure to do so leads to eventual withdrawal of capital which has many consequences including contraction, re-structuring, consolidation, bankruptcy or closure. It is therefore necessary to have good measures of equity risk for managers, planners, and investors. The cost of equity is important in valuing new investment opportunities and evaluating the ongoing performance of existing business projects. This is especially true in the capital intensive energy industry. The purpose of this paper is to calculate the cost of equity in the energy industry, using several different risk measurers, and to compare it with the cost of equity in other industries. The risk measures include systematic risk, total risk, downside risk, and value at risk. The empirical results indicate that while different risk measures produce different values for the cost of equity, the oil and gas industry is no more risky than many other industries. The energy equipment and services industry is, however, riskier than other industries. These results are useful for managers, budget planners, policy makers, and investors who are interested in the energy industry.

This paper is organized as follows. Section 2 presents the empirical approach while section 3 discusses the data. Various risk measures are presented and discussed in section 4 while section 5 reports estimates for the cost of equity. Concluding remarks are contained in section 6.

2. Empirical Approach

This section describes the empirical approach used to calculate the various risk measures and associated required returns. The starting point is to recall that any required return consists of a risk free rate and a risk premium.

$$(1)\ RR_i = R_f + (RP_w)(RM_i)$$

In equation (1), RR_i is the required return on equity i, R_f is the risk free rate, RP_w is the world market risk premium, and RM_i is a risk measure for equity i. Several risk measures are considered and studied. Systematic risk (SR) is measured by the capital asset pricing model (CAPM) beta. Systematic risk has a long history and is one of the most widely used measures of risk (Campbell, Low, and Mackinlay (1997)). Total risk (TR) is measured by the standard deviation of stock returns. Total risk is also widely used and is particularly appropriate in segmented markets. Value at risk (VaR) is a well known measure of the expected losses in extreme downturns (Alexander (2001)). VaR is the loss that is expected to be exceeded with a 5% probability during the next time period. The greater the value of VaR, the greater the potential loss within the defined time period for a particular confidence interval. It is usual to assume a 95% confidence interval and a time period of one month. In this case, VaR is the loss that is expected to be exceeded with a 5% probability during the next month. For example, assume that the United States stock market (measured by the S&P 500) has a VaR of $8. This means that for every $100 invested in the United States stock market, there is a 5% chance of losing $8 or more in any given month. Equation (2) specifies the VaR calculation.

$$(2)\ VaR = (1 - \exp(-1.645\sigma))*100$$

In equation (2), 1.645 is the critical value for a 95% confidence interval and σ is the monthly standard deviation of the monthly continuously compounded total returns.

Another perhaps lesser known measure of risk is downside risk. Downside risk (DR) is measured using semi standard deviation of returns.

$$(3)\ \Sigma_B = \sqrt{(1/T)\sum_{t=1}^{T}(R_t - B)^2} \quad \text{for all } R_t < B$$

In equation (3), R_t are monthly returns and B is the benchmark. There are a number of different benchmark returns and consequently several different measures of downside risk. A popular choice for B is the arithmetic mean (μ) of the continuously compounded monthly returns. In this, case equation (3) reads Σ_μ. Other measures for the benchmark include the risk free rate, and zero (Estrada (2000), Harvey (2000)).

3. Data

The data used in this study consist of monthly data on 24 global industry total returns over the period January 1996 to December 2000. The total return series are expressed in U.S. dollars. The industry data come from Morgan Stanley Capital International (MSCI) and include the energy industry. Total returns on the MSCI world index are also included in the data set. The MSCI world energy sector consists of two industries, energy equipment and services, and oil and gas. Both of these industries are included in the data set. Since the energy industry is a global industry it is useful to make comparisons with other global industries. The MSCI database is ideally suited for doing this.

4. Risk Measures

Monthly statistics for the 24 industries are reported in Table 1. Mean monthly continuously compounded returns vary from a low of –0.02% for real estate to a high of 1.92% for technology hardware. Some industries like energy equipment and services and software services have very high standard deviations. The information ratio (μ/σ) measures returns relative to risk. Risk adverse individuals prefer high values of the information ratio. This indicates that industries like pharmaceuticals and biotechnology, food and drug, utilities, and insurance are particularly attractive because these industries have high return to risk ratios. The energy equipment and services industry has an information ratio below the average for the twenty-four industries while the oil and gas industry has an information ratio above the average for the twenty-four industries. Based on these information ratios, an investment in the oil and gas industry is preferred to an investment in the equipment and services industry. The variable ρ measures the correlation between monthly industry total returns and monthly total world returns. Industries like capital goods, diversified financials and automobiles and components have returns that are highly correlated with world returns. Energy equipment and services and oil and gas have correlation values below the industry average. The final column in Table 1 reports industry beta values. Industry beta values are calculated from a single factor capital asset pricing model.

$$(4)\quad R_{it} = \alpha_i + \beta_i R_{mt} + u_{it}$$

where R_{it} is the industry return, R_{mt} is the return on the MSCI world index, and u_{it} is the residual. Systematic risk (SR) is the beta, β_i, and idiosyncratic risk (IR) is the standard deviation of the residual u_{it}. All of the industry beta values are statistically significant at the 1% level of significance and almost half are larger than unity (Table 1). As expected, software and services and technology hardware have the highest beta values. The energy equipment and services industry has a beta value above the industry average, while the oil and gas industry has a beta value below the industry average. The single factor model in equation (4) can also be used to calculate two downside beta measures (Harvey (2000)). The first measure, downβ_1, is calculated as the coefficient on the market return using observations when industry returns and world market returns are both negative. The second measure, downβ_2, is calculated as the coefficient on the market returns when world market returns are negative.

Table 1. Summary Statistics (monthly values).

industry	μ	σ	information	ρ	β
energy equipment & services	1.13	10.64	0.11	0.45	1.15*
oil & gas	1.19	4.95	0.24	0.53	0.63*
materials	0.05	5.35	0.01	0.73	0.93*
capital goods	0.81	4.61	0.18	0.91	1.00*
commercial services	0.35	4.54	0.08	0.80	0.86*
transportation	0.09	4.00	0.02	0.72	0.69*
automobiles & components	0.35	5.61	0.06	0.84	1.12*
consumer durables	0.29	5.41	0.05	0.82	1.06*
hotels and restaurants	0.31	4.86	0.06	0.73	0.84*
media	1.08	5.31	0.20	0.80	1.01*
retailing	1.18	5.52	0.21	0.73	0.96*
food and drug	1.19	3.97	0.30	0.38	0.36*
beverage and tobacco	0.72	4.46	0.16	0.44	0.47*
personal products	1.13	5.94	0.19	0.51	0.72*
health care equipment	0.45	5.56	0.08	0.57	0.75*
pharmaceuticals & biotechnology	1.66	4.60	0.36	0.45	0.49*
banks	0.64	6.19	0.10	0.79	1.16*
diversified financials	1.38	6.23	0.22	0.86	1.27*
Insurance	1.40	5.35	0.26	0.74	0.94*
real estate	-0.02	6.88	0.00	0.65	1.07*
software and services	1.86	10.18	0.18	0.71	1.73*
technology hardware	1.92	8.75	0.22	0.80	1.67*
telecommunications services	0.91	5.49	0.17	0.71	0.93*
utilities	0.81	2.96	0.27	0.41	0.29*
average	0.87	5.72	0.16	0.67	0.92
world	0.94	4.19	0.22	1.00	1.00

Notes: The information variable is calculated as μ/σ. * indicates a beta value is statistically significant at the 1% level.

For each industry, eleven risk variables are calculated (Table 2). The risk variables are systematic risk (SR) measured be beta, total risk (TR) measured by the standard deviation of returns, idiosyncratic risk (IR) measured by the standard deviation of the residuals from the CAPM model, downside risk (DR) measured by the semi standard deviation of returns, using three different benchmarks, value at risk (VaR), two downside beta measures, skewness, and kurtosis. Two industries, the energy equipment and services industry and the software and services industry stand out as having the largest values of TR, IR, Σ_μ, Σ_f, Σ_0 , VaR, downβ_1 and downβ_2. The -16.05 VaR value of the energy equipment and services industry means that for every $100 invested in this industry there is a 5% probability of losing $16.05 or more in any given month. Based on VaR this industry is more risky than either software and services or technology hardware (two industries that are often singled out for being particularly risky to invest in). By comparison, the oil and gas industry has one of the lowest VaR values at -7.82 and this value is below the industry average. The results in Table 2 show that the oil and gas industry is one of the least risky industries, while the energy equipment and services industry is one of the most risky industries.

Table 2. Risk Variables (monthly values).

industry	SR	TR	IR	Σ_μ	VaR
energy equipment & services	1.15	10.64	9.49	7.96	-16.05
oil & gas	0.63	4.95	4.20	3.32	-7.82
materials	0.93	5.35	3.64	3.73	-8.42
capital goods	1.00	4.61	1.93	3.30	-7.30
commercial services	0.86	4.54	2.75	3.46	-7.20
transportation	0.69	4.00	2.76	2.94	-6.36
automobiles & components	1.12	5.61	3.06	4.14	-8.82
consumer durables	1.06	5.41	3.09	3.80	-8.51
hotels and restaurants	0.84	4.86	3.33	3.63	-7.68
media	1.01	5.31	3.18	4.15	-8.37
retailing	0.96	5.52	3.79	4.11	-8.67
food and drug	0.36	3.97	3.67	3.06	-6.32
beverage and tobacco	0.47	4.46	4.00	3.31	-7.07
personal products	0.72	5.94	5.11	4.42	-9.31
health care equipment	0.75	5.56	4.57	4.40	-8.73
pharmaceuticals & biotechnology	0.49	4.60	4.11	3.46	-7.29
banks	1.16	6.19	3.83	4.66	-9.69
diversified financials	1.27	6.23	3.23	4.79	-9.74
insurance	0.94	5.35	3.62	3.98	-8.43
real estate	1.07	6.88	5.21	4.92	-10.70
software and services	1.73	10.18	7.14	7.47	-15.42
technology hardware	1.67	8.75	5.27	6.62	-13.41
telecommunications services	0.93	5.49	3.84	4.16	-8.63
utilities	0.29	2.96	2.70	2.00	-4.76
average	0.92	5.72	4.06	4.24	-8.95
World	1.00	4.19		3.25	-6.67

Table 2 (continued). Risk Variables (monthly values)

industry	Σ_f	Σ_0	downβ_1	down β_2	skewness	kurtosis
energy equipment & services	10.07	7.42	2.07	1.86	-0.53	1.61
oil & gas	5.87	2.68	0.69	0.90	0.34	0.64
materials	7.01	3.70	1.06	1.31	0.10	1.56
capital goods	6.06	2.90	1.00	1.34	-0.32	1.31
commercial services	6.43	3.29	0.95	1.19	-0.75	1.27
transportation	6.25	2.89	0.86	0.91	-0.53	2.20
automobiles & components	7.04	3.96	1.26	1.53	-0.35	0.77
consumer durables	6.91	3.64	1.14	1.49	0.06	0.74
hotels and restaurants	6.61	3.48	1.09	1.21	-0.73	2.40
media	6.38	3.68	1.18	1.48	-0.95	2.82
retailing	6.45	3.51	0.91	1.14	-0.47	0.45
food and drug	5.44	2.51	0.56	0.46	-0.82	1.14
beverage and tobacco	6.07	2.94	0.88	0.73	-0.38	0.15
personal products	6.65	3.90	0.96	0.96	-0.17	1.78
health care equipment	7.07	4.20	1.34	1.14	-1.20	2.55
pharmaceuticals & biotechnology	5.52	2.67	0.71	0.68	-0.52	-0.12
banks	7.28	4.36	1.37	1.70	-0.85	3.09
diversified financials	6.84	4.20	1.36	1.82	-1.19	4.72
Insurance	6.12	3.36	1.20	1.33	-0.64	3.09
real estate	8.07	4.94	1.16	1.23	-0.26	2.20
software and services	9.27	6.51	1.79	2.52	-0.34	0.26
technology hardware	8.38	5.64	1.56	2.36	-0.59	0.28
telecommunications services	6.65	3.72	0.98	1.36	-0.60	0.37
utilities	5.10	1.55	0.31	0.37	0.20	-0.47
average	6.81	3.82	1.10	1.29	-0.48	1.45
world	5.55	2.82			-0.93	1.79

Notes: The risk variables are systematic risk (SR), total risk (TR), idiosyncratic risk (IR), downside risk Σ_μ , Σ_f , Σ_0, down β_1, down β_2, skewness, kurtosis, and value at risk (VaR).

A correlation matrix between mean returns (MR) and the eleven risk variables indicate that market returns are most highly correlated with Σ_μ, total risk, and VaR (Table 3). Market returns are least correlated with downside risk measure Σ_f. Total risk and downside risk are very highly correlated with each other and downside risk and VaR are very highly correlated with each other. Total risk is perfectly correlated with VaR.

Table 3. Correlation Matrix

	MR	TR	SR	IR	Σ_μ	VaR	Σ_f	Σ_0	SKEW	KURT	downβ_1	downβ_2
MR	1.00	0.40	0.28	0.37	0.42	-0.40	0.13	0.24	-0.15	-0.18	0.20	0.32
TR	0.40	1.00	0.80	0.87	0.99	-1.00	0.96	0.98	-0.10	0.05	0.90	0.84
SR	0.28	0.80	1.00	0.39	0.79	-0.80	0.78	0.79	-0.19	0.21	0.83	0.98
IR	0.37	0.87	0.39	1.00	0.86	-0.86	0.82	0.84	0.01	-0.09	0.68	0.96
Σ_μ	0.42	0.99	0.79	0.86	1.00	-0.99	0.95	0.98	-0.20	0.10	0.92	0.84
VaR	-0.40	-1.00	-0.80	-0.86	-0.99	1.00	-0.96	-0.98	0.11	-0.06	-0.91	-0.84
Σ_f	0.13	0.96	0.78	0.82	0.95	-0.96	1.00	0.99	-0.10	0.10	0.92	0.81
Σ_0	0.24	0.98	0.79	0.84	0.98	-0.98	0.99	1.00	-0.20	0.16	0.94	0.83
SKEW	-0.15	-0.10	-0.19	0.01	-0.20	0.11	-0.10	-0.20	1.00	-0.60	-0.33	-0.22
KURT	-0.18	0.05	0.21	-0.09	0.10	-0.06	0.10	0.16	-0.60	1.00	0.32	0.22
downβ_1	0.20	0.90	0.83	0.68	0.92	-0.91	0.92	0.94	-0.33	0.32	1.00	0.88
downβ_2	0.32	0.84	0.98	0.46	0.84	-0.84	0.81	0.83	-0.22	0.22	0.88	1.00

Notes: The variables are market return (MR), systematic risk (SR), total risk (TR), idiosyncratic risk (IR), downside risk measures Σ_μ, Σ_f , Σ_0 , value at risk (VaR), skewness, kurtosis, and downside beta measures downβ_1 and downβ_2

A cross section regression analysis is used to investigate the relationship between mean returns and various risk variables (Table 4). The lack of explanatory power for the systematic risk variable is evident as the estimated slope coefficient is statistically insignificant at conventional levels of significance and the R^2 value is one half of the R^2 value from the total risk regression. The evidence suggests that systematic risk is not significantly related to market returns.

Table 4. Cross Section Regression Analysis

$MR_i = \alpha_0 + \alpha_1 RV_i + u_i$

RV	α_0	α_1	R^2	Adjusted R^2	RESET
SR	0.47	0.44	0.08	0.03	0.02
	0.19	0.24			
TR	0.16	0.12	0.16	0.13	0.61
	0.63	0.04			
IR	0.33	0.13	0.14	0.10	0.47
	0.23	0.07			
Σ_μ	0.14	0.17	0.18	0.14	0.53
	0.63	0.03			
VaR	0.12	-0.08	0.16	0.12	0.62
	0.74	0.04			
Σ_f	0.45	0.06	0.02	-0.03	0.09
	0.53	0.57			
Σ_0	0.46	0.11	0.06	0.02	0.33
	0.17	0.24			
downβ_1	0.54	0.30	0.04	-0.00	0.21
	0.10	0.30			
downβ_2	0.43	0.34	0.10	0.06	0.02
	0.20	0.16			
Skewness	0.77	-0.21	0.02	-0.02	0.61
	0.00	0.43			
Kurtosis	0.99	-0.08	0.03	-0.01	0.05
	0.00	0.45			

Notes: White's (1980) heteroskedasticity corrected probability values reported below coefficient estimates. RESET shows the probability values for a Ramsey (1969) regression specification test with 2 degrees of freedom.

There are several plausible explanations for why the empirical evidence suggests that systematic risk is not significantly related to market returns The CAPM beta measures incremental systematic risk in a perfectly diversified portfolio. The CAPM is built upon a set of fairly restrictive assumptions. A violation of one or more of these assumptions can lead to biased estimates of beta. For example, the beta may not be constant through time. The alpha may not be constant through time. The error variance may not be constant through time. The errors may be correlated through time. Returns may be non-linearly related to market returns.

The returns of the portfolio and the risk free rate may be measured incorrectly or related in a non-linear way. There may be other sources of risk. The correlation between the market return and the errors may not be zero. The world market portfolio is not mean variance efficient. The regression specification test (Ramsey (1969)) indicates some evidence of regression miss-specification in the systematic risk equation. This indicates that there are missing risk factors.

Several risk measures (Σ_f , downβ_1 , skewness, and kurtosis) exhibit a profound lack of explanatory power as indicated by the low R^2 values. These R^2 values range from 2% to 4%. In addition these risk measures even have negative adjusted R^2 values.

In contrast, four other risk variables, TR, IR, Σ_μ, and VaR are significantly related to market returns. The R^2 values for these regression models range from 14% to 18%. The regression specification tests indicate that none of these models show evidence of regression miss-specification. The regression model using downside risk (mean) as a risk variable has the highest R^2 value indicating downside risk can explain 18% of the variation in industry stock returns. This is impressive given how difficult it is to model stock returns.

5. THE COST OF EQUITY

The cost of equity is important in valuing new investment opportunities and evaluating the ongoing performance of existing business projects. From equation (1) any required return consists of a risk free rate and a risk premium.

$$(5)\quad RR_i = R_f + (RP_w)(RM_i)$$

In equation (5), RR_i is the required return on equity i, R_f is the risk free rate, RP_w is the world market risk premium, and RM_i is a risk measure for equity i. This section focuses on three risk measures based on systematic risk (RM_{SR}), total risk (RM_{TR}), and downside risk (RM_{DR}) calculated from Σ_μ. Total risk and downside risk are included because they have the best fit in the regression models (Table 4). Systematic risk is included because it is so well known. The risk measures and associated cost of equity (CE) are calculated as follows.

$$(6)\quad RM_{SR} = \beta_i / \beta_w = \beta_i$$

$$CE_{SR,i} = R_f + (RP_w)\,\beta_i$$

$$(7)\quad RM_{TR} = \sigma_i / \sigma_w$$

$$CE_{TR,i} = R_f + (RP_w)\,\sigma_i / \sigma_w$$

$$(8)\quad RM_{DR} = \Sigma_{\mu i} / \Sigma_{\mu w}$$

$$CE_{DR,i} = R_f + (RP_w)\,\Sigma_{\mu i} / \Sigma_{\mu w}$$

Table 5 reports annualized means, standard deviations, risk measures, and cost of equity values for the 24 global industries. For comparison purposes, the annualized return on crude oil prices (using the West Texas intermediate continuous crude oil futures contract on the New York Mercantile Exchange) was 13.60% with an associated standard deviation of 27.92%. This yields an information ratio of 0.49. As expected, the information ratios reported in Table 5 yield the same qualitative information as those ratios reported in Table.1 On a risk adjusted basis, industries like pharmaceuticals and biotechnology, food and drug, utilities, and insurance are particularly attractive because these industries have high return to risk ratios. The energy equipment and services industry has an information ratio below the industry average while the oil and gas industry has an information ratio above the industry average. Based on these information ratios, an investment in the oil and gas industry is preferred to an investment in the equipment and services industry. The results reported in Table 5 show that risk measures based on total risk or downside risk are much larger than risk measures based on systematic risk In some industries, total risk and downside risk are more than twice as large as the systematic risk measures. Clearly these results have implications for the returns that investors expect to earn.

In order to calculate the cost of equity for each industry, numerical values for a risk free rate and world risk premium are required. In the calculations of the cost of equity, a risk free rate of 5% was used as this was the average three-month U.S. Treasury bill rate over the period January 1996 to December 2000. A world risk premium of 5.5% was used. This was the same value used in Estrada (2000) and is reasonably accurate for the time period under consideration. The cost of equity based on systematic risk is the lowest of the three cost of equity measures. For most industries the cost of equity ranking is $CE_{TR} > CE_{DR} > CE_{SR}$. The differences between these three risk measures can be striking. In the energy equipment and services industry CE_{SR} is 11.32% per year while CE_{TR} and CE_{DR} are 18.95% and 18.46% per year respectively. This means that an investor in the energy equipment and services industry would be expecting to earn 11.32% per year if the cost of equity was calculated using systematic risk. In comparison, this investor would expect to earn 18.95% per year if the cost of equity was calculated from a total risk measure. This is a substantial difference in risk adjusted expected returns. In the oil and gas industry CE_{SR} is 8.45% per year while CE_{TR} and CE_{DR} are 11.49% and 10.62% per year respectively. The software and services industry and technology hardware industries has the highest cost of equity based on systematic risk. The energy equipment and services industry has the highest cost of equity based on either total risk or downside risk. The oil and gas industry has cost of equity values based on systematic risk, total risk, and downside risk that are below the industry averages.

Table 5. Risk Measures and the Cost of Equity (annualized returns)

Industry	μ	σ	info	Σ_{μ}	RM_{SR}	RM_{TR}	RM_{DR}	CE_{SR}	CE_{TR}	CE_{DR}
energy equipment & services	13.55	36.85	0.37	27.56	1.15	2.54	2.45	11.32	18.95	18.46
oil & gas	14.28	17.15	0.83	11.5	0.63	1.18	1.02	8.45	11.49	10.62
materials	0.61	18.53	0.03	12.92	0.93	1.28	1.15	10.14	12.02	11.31
capital goods	9.69	15.97	0.61	11.44	1	1.1	1.02	10.49	11.05	10.59
commercial services	4.2	15.74	0.27	11.99	0.86	1.08	1.07	9.74	10.96	10.86
transportation	1.1	13.84	0.08	10.18	0.69	0.95	0.9	8.79	10.24	9.97
automobiles & components	4.23	19.44	0.22	14.34	1.12	1.34	1.27	11.17	12.36	12
consumer durables	3.46	18.73	0.18	13.15	1.06	1.29	1.17	10.82	12.09	11.42
hotels and restaurants	3.73	16.82	0.22	12.56	0.84	1.16	1.12	9.63	11.37	11.13
media	12.94	18.4	0.7	14.36	1.01	1.27	1.28	10.58	11.97	12.01
retailing	14.16	19.11	0.74	14.23	0.96	1.32	1.26	10.25	12.23	11.95
food and drug	14.22	13.75	1.03	10.6	0.36	0.95	0.94	7	10.21	10.18
beverage and tobacco	8.68	15.45	0.56	11.47	0.47	1.06	1.02	7.59	10.85	10.6
personal products	13.6	20.58	0.66	15.32	0.72	1.42	1.36	8.97	12.79	12.48
health care equipment	5.36	19.25	0.28	15.25	0.75	1.33	1.35	9.15	12.29	12.45
pharmaceuticals & biotechnology	19.88	15.93	1.25	11.99	0.49	1.1	1.07	7.71	11.03	10.86
banks	7.65	21.46	0.36	16.15	1.16	1.48	1.43	11.39	13.13	12.89
diversified financials	16.61	21.59	0.77	16.61	1.27	1.49	1.48	11.99	13.18	13.11
insurance	16.76	18.54	0.9	13.79	0.94	1.28	1.23	10.18	12.02	11.74
real estate	-0.22	23.83	-0.01	17.06	1.07	1.64	1.52	10.9	14.02	13.34
software and services	22.29	35.27	0.63	25.87	1.73	2.43	2.3	14.52	18.36	17.64
technology hardware	23.01	30.31	0.76	22.93	1.67	2.09	2.04	14.16	16.48	16.2
telecommunications services	10.88	19.01	0.57	14.42	0.93	1.31	1.28	10.14	12.2	12.04
utilities	9.68	10.26	0.94	6.93	0.29	0.71	0.62	6.6	8.89	8.38
average	10.43	19.83	0.54	14.69	0.92	1.37	1.31	10.07	12.51	12.18
world	11.27	14.53	0.78	11.26	1	1	1	10.5	10.5	10.5

While systematic risk is a very well known and widely used measure of market risk, the results in this paper suggest that a prudent investor might also wish to calculate risk measures based on total risk and downside risk. For example, the cost of equity based on systematic risk is 11.32% per year in the energy equipment and services industry and 10.82% in the consumer durables industry. Clearly the energy equipment and services industry is much more risky than the consumer durables industry and, while the risk is not borne out by the systematic risk calculation, it is evident in the cost of equity calculation based on total risk and downside risk. For the consumer durables industry, the three cost of equity measures produce very similar results from a low of 10.82% (based on systematic risk) to a high of 12.09% (based on total risk). For industries like capital goods, commercial services, automobiles and parts, consumer durables, and diversified financials the choice of which risk measure to use is not that important because the three cost of equity measures are fairly similar (less than a 1.5% difference between the largest and lowest cost of equity values) for each of these industries. For other industries like energy equipment and services and software and services, the difference between the three cost of equity measures can be large (as much as 7.6% per year for the energy equipment and services industry).

6. Conclusions

This paper uses some recent developments in financial modeling to calculate eleven risk measures and three cost of equity measures for the global energy industry. In addition to using well know risk measure like systematic risk, and total risk, this paper includes some other more recent measures of risk like downside risk, and value at risk. While different cost of equity measures produce different numerical results, the qualitative results are fairly similar suggesting that the energy equipment and services industry is much more risky than other industries, while the oil and gas industry is less risky than other industries. These results are useful for managers, budget planners, investors, and policy makers who need to have risk measures and cost of equity measures for the global energy industry. For individual companies, these cost of equity measures provide a useful benchmark with which to compare a firm's cost of equity with the energy industry cost of equity. These cost of equity measures are useful to compare the energy industry cost of equity with the cost of equity from twenty-three other industries.

These results indicate that investors should expect annualized returns in the energy equipment and services industry that are comparable to those in the software and services industry. The oil and gas industry is much less risky than the energy equipment and services industry and investors in oil and gas stocks should expect annualized returns below the average for the twenty-four industries. While systematic risk is a very well known and widely used measure of market risk, the results in this paper suggest that a prudent investor, manager, budget planner, or policy maker might also wish to calculate risk measures based on total risk and downside risk.

REFERENCES

Alexander, C. 2001. Market Models: *A Guide to Financial Data Analysis*, John Wiley & Sons.

Campbell, J.Y., A.W. Low, and A.C. Mackinlay 1997. *The Econometrics of Financial Markets*, Princeton University Press.

Estrada, J. 2000. The cost of equity in emerging markets: A downside risk approach, *Emerging Markets Quarterly*, 4,3, 19-30.

Fusaro, P.C. 1998. *Energy Risk Management: Hedging Strategies and Instruments for the International Energy Markets*. McGraw Hill, New York.

Ghemawat, P. and F. Ghadar 2000. The dubious logic of global megamergers, Harvard Business Review, 78, 4, 65-74.

Harvey, C. R. 2000. Drivers of expected returns in international markets, Emerging Markets Quarterly, 4,3, 32-48.

Ramsey, J. 1969. Tests for specification errors in classical linear least squares regression analysis, Journal of the Royal Statistical Society B, 31, 350-371.

Sadorsky, P. 2001. Risk factors in stock returns of Canadian oil and gas companies. Energy Economics, 23, 17-28.

Sadorsky, P. 2002. Time-Varying Risk Premiums in Petroleum Futures Prices. Energy Economics, forthcoming.

Stern Stewart 2001. One way to close the oil industry's credibility gap: Hedge PDP production and buy back stock, EVAluation, 3, 6

White, H. 1980. A heteroskedasticity-consistent covariance matrix estimator and a direct test for heteroskedasticity, Econometrica, 48, 817-38.

INDEX